Agile Strategy Management in the Digital Age

David Wiraeus • James Creelman

Agile Strategy Management in the Digital Age

How Dynamic Balanced Scorecards Transform Decision Making, Speed and Effectiveness

palgrave
macmillan

David Wiraeus
Stratecute Group
Gothenburg, Sweden

James Creelman
Creelman Strategy Alliance
London, UK

ISBN 978-3-030-09460-7   ISBN 978-3-319-76309-5   (eBook)
https://doi.org/10.1007/978-3-319-76309-5

© The Editor(s) (if applicable) and The Author(s) 2019
Softcover re-print of the Hardcover 1st edition 2019
This work is subject to copyright. All rights are solely and exclusively licensed by the Publisher, whether the whole or part of the material is concerned, specifically the rights of translation, reprinting, reuse of illustrations, recitation, broadcasting, reproduction on microfilms or in any other physical way, and transmission or information storage and retrieval, electronic adaptation, computer software, or by similar or dissimilar methodology now known or hereafter developed.
The use of general descriptive names, registered names, trademarks, service marks, etc. in this publication does not imply, even in the absence of a specific statement, that such names are exempt from the relevant protective laws and regulations and therefore free for general use.
The publisher, the authors, and the editors are safe to assume that the advice and information in this book are believed to be true and accurate at the date of publication. Neither the publisher nor the authors or the editors give a warranty, express or implied, with respect to the material contained herein or for any errors or omissions that may have been made. The publisher remains neutral with regard to jurisdictional claims in published maps and institutional affiliations.

This Palgrave Macmillan imprint is published by the registered company Springer Nature Switzerland AG
The registered company address is: Gewerbestrasse 11, 6330 Cham, Switzerland

I dedicate this book to my wife Vanja and our child Ella, as well as my sister Frida and parents Ulla and Anders, with all my love.
David Wiraeus
For my great nephews Kian Creelman and Ezra French and my great niece Arrabella French. Enjoy the long journey ahead.
James Creelman

Foreword

More than 25 years have passed since Bob Kaplan and I introduced the concept of the Balanced Scorecard through a Harvard Business Review article. This relayed the findings from a research project we led in 1990 with 12 large companies to find better ways to measure performance, rather than relying solely on financial measures. At that time, we were transitioning into the knowledge age, in which intangible assets were becoming more valuable than tangible ones, and where the increasing speed of change in markets meant that financial measures were no longer reliable predictors of future performance.

Financial results would remain, and continue to be, important, at least for commercial entities, but what were the non-financial drivers of those outcomes? This was the question we grappled with.

The answer proved simple and logical. Customers delivered financial results; the organization had to ensure its internal processes delivered value to the customer and that they possessed the required skills and capabilities to deliver those processes effectively and efficiently.

These observations were translated into the Balanced Scorecard framework, which comprised *Financial*, *Customer*, *Internal Process*, and *Learning and Growth* perspectives, each of which contained objectives (what we want to achieve), measures and targets (how we will monitor progress), and initiatives (how we will deliver to those targets).

A further question we wrestled with was why 90% of organizations failed to deliver to their strategy, even when it was well thought-out and logical. We found that the Balanced Scorecard could describe and operationalize strategies that previously were generally restricted to a very detailed strategic plan, which rarely left the boardroom shelf. Bob and I chronicled the successes of the original tranche of scorecard users in our first book, *The Balanced Scorecard: Translating Strategy into Action*.

In our continued research, we found that some of the early Balanced Scorecard users, such as Mobil Oil's North American Division, gained additional value when the strategic objectives were laid out separately to show the causal effect from the learning and growth perspective, through internal processes to customer and financial. Furthermore, although originally launched to overcome strategic performance management and measurement challenges in commercial organizations, government and not-for-profit entities, such as the City of Charlotte, North Carolina, soon adopted the framework. However, to meet their needs, such organizations reordered the perspectives, with stakeholder at the top (typically replacing the term customer) and financial lower down the Strategy Map.

These first Strategy Maps proved as valuable to users as the original Balanced Scorecard itself, as we explained in our second book, *The Strategy-Focused Organization: How Balanced Scorecard Companies Thrive in the New Business* Environment and described fully in our third book, *Strategy Maps: Converting Intangible Assets into Tangible Outcomes*.

The story did not end there. We continued to learn from the experiences of an ever-growing number of users. Our fourth book, *Alignment: Using the Balanced Scorecard to Create Corporate Synergies*, documented the value organizations gained from cascading the Balanced Scorecard from the corporate level to business units and then to operating departments and support functions, as well as being the basis for strategically aligning external stakeholders.

Our final book. *The Execution Premium: Linking Strategy to Operations for Competitive Advantage*, set out to offer a complete strategy management system through a six-stage model: defining the strategy, translating the strategy, aligning the organization, aligning operations, monitoring and learning, and testing and adapting.

However, even completing the fifth and final book does not mean the end of the story. On introducing the Balanced Scorecard framework and methodology, Bob Kaplan and I realized we were launching a revolution, not a static system. We knew it would continue to evolve to meet the strategic requirements of organizations in ever-changing and fast-moving markets.

Our work, and most notably the final book, serves as the inspiration for this book, *Agile Strategy Management in the Digital Age – How Dynamic Balanced Scorecards Transform Decision Making, Speed, and Effectiveness*.

Just as Bob and I grappled with the challenges of transitioning from the industrial age to the knowledge age, Wiraeus and Creelman turn their attention to the challenges of moving into the digital age.

Changing roles from Balanced Scorecard historian to Balanced Scorecard futurist, the authors inventory the issues that must be integrated into the management systems of the future – They are to be commended for the audacity of their undertaking and for the reach of their results.

The revolution continues.

Massachusetts
March, 2018

David P. Norton

Acknowledgements

This book could not have been written without the advice, knowledge, and support of many people, whom we here acknowledge.

Bill Barberg (Insightformation), James Bass (Certified Scrum Professional), Bjarte Bogsnes (Statoil), James Coffey (Beyond Scorecard), Deepanjan Chakrabarty (previously Palladium), Marcello Coluccia (Imerys Graphite & Carbon), Jade Evans (Palladium), Liam Fahey (Leadership Inc.), Elena Gómez Domenech (Palladium) Mihai Ionescu (Strategys), Saliha Ismail (Ministry of Works, Municipality Affairs & Urban Planning, Bahrain), Brett Knowles (pm2Consulting), Stanley Labovitz (SurveyTelligence), Armen Mnatsakanyan (ConconFM), Sandy Richardson (Collaborative Strategy), Hubert Saint-Onge (Saint-Onge Alliance), Alistair Schneider (startupsinnovation.com), Andreas de Vries (Oil & Gas Strategy Management Speuncialist expert), Iain Wicking (Oyonix Group).

We also extend our gratitude to Dr. David Norton for providing the Foreword to this book and to all our ex-colleagues in Palladium, who shared their knowledge and experience with us over many years.

Thanks also to Palgrave Macmillan's Stephen Partridge and Gabriel Everington for their continued support and guidance.

Finally, James would like to thank Matt Sabbath Stark, Hugh Sturrock, Hugh Macleod, and Anto Brownes for their many nights in the Earl of Derby, Kilburn, London, patiently listening to his constant talking about the writing of the book.

Contents

1 **Digital Age Strategy Management: From Planning to Dynamic Decision Making**   1
   Introduction   1
   No "Perfect" Management Solution   1
   It's All About Evolution   2
   Common Challenges   5
   The Scourge of Silo-Based Working   6
   The Strategy Function and Process   8
   Assumptions that Must Be Verified in Execution   10
   End-to-End Process Management   10
   Strategic Innovation   12
   The Importance of Agility   13
   An Agile and Adaptive Model for Strategy Execution in the Digital Age   14
   Parting Words: Shifting Paradigms   19
   Self-Assessment Checklist   20
   References   21

2 **From Industrial- to Digital-Age-Based Strategies**   23
   Introduction   23
   Challenging the Notion that "Strategy is dead!"   25
   Defining Strategy   26
   Defining the Sense of Purpose   28
   Finance-Based Planning   34

Capturing the Voice of the Customer	36
From Finance-Based to Technology-Based Planning	
Self-Assessment Checklist	42
References	44

3 Agile Strategy Setting — 45

Introduction	45
Crafting a Vision Statement	45
Identifying the Value Gap	48
Environmental Scanning	49
SCOPE Situational Analysis	53
The Danger of Being Frozen in Time	54
Senior Management Interviews	55
Using an External Facilitator	56
A Strategic Change Agenda	59
Parting Words	62
Self-Assessment Checklist	65
References	66

4 Strategy Mapping in Disruptive Times — 69

Introduction	69
Starting with the Strategy Map	69
Writing Objectives	70
Keeping Strategy Maps Focused	73
Objective Statements	73
The Value of Strategic Themes	75
The Power of Cause and Effect	79
Parting Words	83
Self-Assessment Checklist	87
References	88

5 How to Build an Agile and Adaptive Balanced Scorecard — 89

Introduction	89
The Purpose of KPIs	90
Four Steps of KPI Selection	92
The Balanced Scorecard Is Not a Measurement System	96
The Science of Measurement	98
Setting Targets	103
Choosing Strategic Initiatives	106

Parting Words	109
Self-Assessment Checklist	110
References	112

6 Driving Rapid Enterprise Alignment — 113
Introduction	113
Traditional Approaches	113
An Agile Approach	117
Team Discussions	120
Alignment and Synergies	122
Parting Words	125
Self-Assessment Checklist	126
References	127

7 Aligning the Financial and Operational Drivers of Strategic Success — 129
Introduction	129
Aligning Budgeting with Strategy	129
Killing the Budget	134
Linking Operations to Strategy	139
Parting Words	147
Self-Assessment Checklist	148
References	149

8 Developing Strategy-Aligned Project Management Capabilities — 151
Introduction	151
Agile Project Management	153
The Transformation Office	156
Measuring Impact	159
A Portfolio Approach	160
Parting Words	162
Self-Assessment Checklist	165
References	166

9 Unleashing the Power of Analytics for Strategic Learning and Adapting — 167
Introduction	167
Correlations and Causality	168

Analytics and KPIs	169
Analytics and Decision Making	169
Advanced Analytics and Strategy Management	172
Required Capabilities	175
Simple Analytics	177
Performance Reviews	179
Strategy Reports	182
A Decline in the Importance of KPIs	182
The Strategy Refresh	183
Implications for the Strategy Office: Practitioner View	185
Parting Words	186
Self-Assessment Checklist	189
References	190

10 How to Ensure a Strategy-Aligned Leadership — 191

Introduction	191
The Importance of Context	192
Leadership for the Execution of Strategy	193
Strategic Leadership: Research Evidence	199
Agile Leadership in an Age of Digital Disruption	201
Assessing the Models	204
Parting Words	205
Self-Assessment Checklist	205
References	206

11 How to Ensure a Strategy-Aligned Culture — 207

Introduction	207
Defining Culture	209
Corporate Values	210
Leadership and Culture	213
Driving Culture Change with the Balanced Scorecard	214
Cultural Assessment	215
Integrating Data	216
Parting Words	218
Self-Assessment Checklist	222
References	223

12 Ensuring Employee Sense of Purpose in the Digital Age — 225

Introduction	225
Gallup Research Evidence	225

	Changing the Employee-Employer Relationship	227
	The End of Appraisals	228
	The Dangers of Assigning KPIs to Individuals	229
	Changing the Conversation	230
	Theory X and Theory Y	231
	A Sense of Purpose	231
	Deloitte Research Findings	232
	Communication	234
	Human Capital Development to Execute Strategy	237
	Parting Words	238
	Self-Assessment Checklist	240
	References	241
13	**Further Developments: Driving Sustainable Value Through Collaborative Strategy Maps and Scorecards**	**243**
	Introduction	243
	Corporate Social Responsibility	243
	Triple Bottom Line	244
	Nova Nordisk Case Illustration	244
	Sustainability Strategy Map	245
	Shared Value	248
	Shared Value Explained	248
	Positive Impact	249
	Networked Organizations	249
	Case Illustration: Thriving Weld	250
	Robust Shared Measurement System	252
	Improved Engagement and New Actions	254
	Parting Words	254
	Self-Assessment Checklist	256
	References	257
14	**Conclusion and 25 Key Strategic Questions**	**259**
	Introduction	259
	Agile and Adaptive	259
	25 Key Strategic Questions	260
	Stage 1: How to Formulate Strategies for the Digital Age	260
	Stage 2: How to Build an Agile and Adaptive Balanced Scorecard	263
	Stage 3: Driving "Rapid" Enterprise Alignment	265
	Stage 4: Getting Results Through Agile Strategy Execution	266

Stage 5: Unleashing the Power of Analytics for Strategic Learning
and Adapting 268
Underpinning the Model 270
Final Words 272
References 272

Index 273

List of Figures

Fig. 1.1	First generation Balanced Scorecard	2
Fig. 1.2	A Strategy Map. (Source: Palladium)	3
Fig. 1.3	The Execution Premium Process	4
Fig. 1.4	Agile and adaptive strategy execution model	5
Fig. 1.5	Age 2 Strategy Management System	20
Fig. 2.1	Stage 1: How to formulate strategies for the digital age	24
Fig. 2.2	Mapping the customer journey	38
Fig. 3.1	Stage 1: How to formulate strategies for the digital age	46
Fig. 3.2	A PESTEL analysis	50
Fig. 3.3	Porter's Five Forces framework	50
Fig. 3.4	A SWOT analysis template	51
Fig. 3.5	Strategsys SWOT analysis. (Source: Strategsys)	52
Fig. 3.6	Strategic Change Agenda	60
Fig. 3.7	FBI Strategic Change Agenda	61
Fig. 3.8	FBI strategy map	62
Fig. 3.9	Strategic Change Agenda, arranged from four perspectives	63
Fig. 4.1	Stage 2: How to build an "agile," and "adaptive," Balanced Scorecard System	72
Fig. 4.2	Strategic themes within the internal process perspective	76
Fig. 4.3	Example of strategic themes within customer and internal process perspectives. (Source: Palladium)	76
Fig. 4.4	A.W. Rostamani strategic pillars	77
Fig. 4.5	Mobil oil linkage model for 1994	81
Fig. 4.6	Intellectual capital model	82
Fig. 4.7	The risk Bow-Tie	86
Fig. 5.1	Stage 2: How to build an "agile," and "adaptive," Balanced Scorecard System	90
Fig. 5.2	Three sub-processes of hospital example strategic objective	94

Fig. 5.3	Initiative scoring model	108
Fig. 6.1	Stage 3. Driving rapid enterprise alignment	114
Fig. 7.1	Stage 4: getting results through agile and adaptive strategy execution	130
Fig. 7.2	Statoil's alternative to the budget	132
Fig. 7.3	Statoil's menu for managing costs	133
Fig. 7.4	Using rolling forecasts alongside Balanced Scorecard targets and initiatives. (Source: Palladium)	137
Fig. 7.5	Driver models provide the analytical framework to focus on key leverage points and to link operational KPIs and action plans to strategic priorities	137
Fig. 7.6	Decomposing an internal process objective into operational drivers and KPIs	145
Fig. 7.7	Patient access operational dashboard	146
Fig. 8.1	Stage 4: getting results through agile and adaptive strategy execution	152
Fig. 8.2	Execution: the "seventh" stage of the execution premium process	152
Fig. 8.3	The roles of the three offices within a transformation office. (Source: ShiftIn)	159
Fig. 8.4	Initiative realization time-lag. (Source: Synergys)	161
Fig. 9.1	Stage 9. Unleashing the power of analytics for strategic learning and adapting	168
Fig. 9.2	The 4Vs of big data	170
Fig. 9.3	Gartner analytic ascendancy model. (Source: Gartner)	171
Fig. 9.4	A Fishbone diagram	177
Fig. 9.5	The five whys problem solving tool	178
Fig. 10.1	Strategy-aligned leadership and culture	192
Fig. 10.2	The Leadership for the Execution of Strategy Model. (Source: Palladium)	194
Fig. 11.1	Strategy-aligned leadership and culture	208
Fig. 11.2	A Strategy Map with a "build a high performance culture" objective	214
Fig. 11.3	High-level Eggi schematic, showing alignment as the core focus. (Source: SurveyTelligence)	221
Fig. 12.1	Create a strategt-aligned workforce for the 4th industrial revolution	226
Fig. 12.2	The causal relationship between the four human capital pillars and ultimate financial or mission success	237
Fig. 13.1	A Sustainability Strategy Map. (Source: pm2Consulting)	246
Fig. 13.2	A Sustainability Strategy Map aligned to the 21 Sustainability Business Goals. (Source: pm2Consulting)	247
Fig. 13.3	Thriving Weld County Strategy Map. (Source: Insightformation)	251

List of Tables

Table 1.1	Self-assessment checklist	21
Table 2.1	Self-assessment checklist	43
Table 3.1	Self-assessment checklist	66
Table 4.1	Self-assessment checklist	88
Table 5.1	Self-assessment checklist	111
Table 8.1	The difference between traditional and agile approaches to project management	154
Table 8.2	Self-assessment checklist	166
Table 9.1	Performance area breakdown for customers that provided an overall rating of 5	179
Table 9.2	Performance area breakdown for customers that provided an overall rating of 3	179
Table 9.3	Self-assessment checklist	189
Table 10.1	Self-assessment checklist	206
Table 12.1	Self-assessment checklist	241
Table 13.1	Self-assessment checklist	256

1

Digital Age Strategy Management: From Planning to Dynamic Decision Making

> The Plan is Nothing, Planning is everything,
> *General (later President) Dwight Eisenhower*

Introduction

Let's get the clichés out of the way (for now!). We live in turbulent, unpredictable times. Constant disruption is the new status-quo. We need to be more customer-centric. We live in a digitally driven world, where data and ideas move across the globe and at the speed of light and the touch of a button: the digital economy is changing everything. We have the most highly educated workforce in history and the Millennials are fundamentally changing the essence of the employer-employee relationship. We could go on.

Billions of words are being written in books, blogs, and so on, and as many again spoken in podcasts, seminars, conferences, and their like, explaining what all this means in terms of strategic management, future organizational structures, teamwork, innovation, and the rest. Perfect solutions—the much sought-after magic bullets—are being proffered.

No "Perfect" Management Solution

Here's our take: various frameworks, models, and solutions will emerge to capitalize on the opportunities of the digital age (or the now often-called 4th industrial revolution) and to manage the accompanying risks. We cannot, with any degree of accuracy, predict which will be particularly useful. No one has ever accurately predicted how revolutions will play out or what a post-revolutionary world will look like: it is likely that when it comes to how organizations go to market and create value, there will never be such a time as post-revolutionary.

We can be more confident in predicting that there will never be a "perfect" management solution. Rather, there will be a combination of approaches that will work well in a specific context for a specific timeframe, before becoming dysfunctional and no longer appropriate. To add another cliché, we might always be cursed to, "live in interesting times."

It's All About Evolution

Think of it this way: the model we propose builds on the seminal works of many pioneering thinkers. Most notably, the ground-breaking work of Harvard Business Professors Robert Kaplan and Dr. David Norton in evolving the Balanced Scorecard Strategy Execution System. The Balanced Scorecard was introduced as essentially a measurement system that addressed the issues of being overly reliant on financial metrics for managing organizations by adding the non-financial balance (Fig. 1.1 shows the first generation Balanced Scorecard) [1]. The next step saw the inclusion of Strategy Maps to capture causal relationships between the non-financial drivers and the financial outcomes (Fig. 1.2) and, finally, to the articulation of the Execution

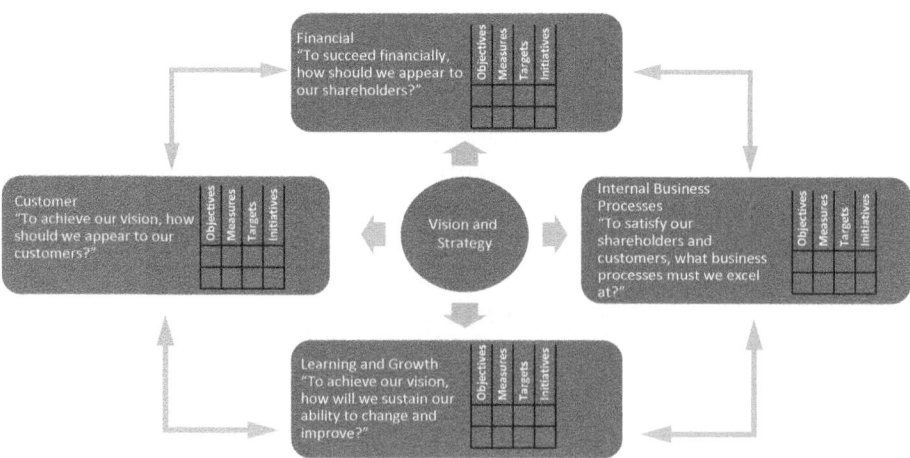

Fig. 1.1 First generation Balanced Scorecard

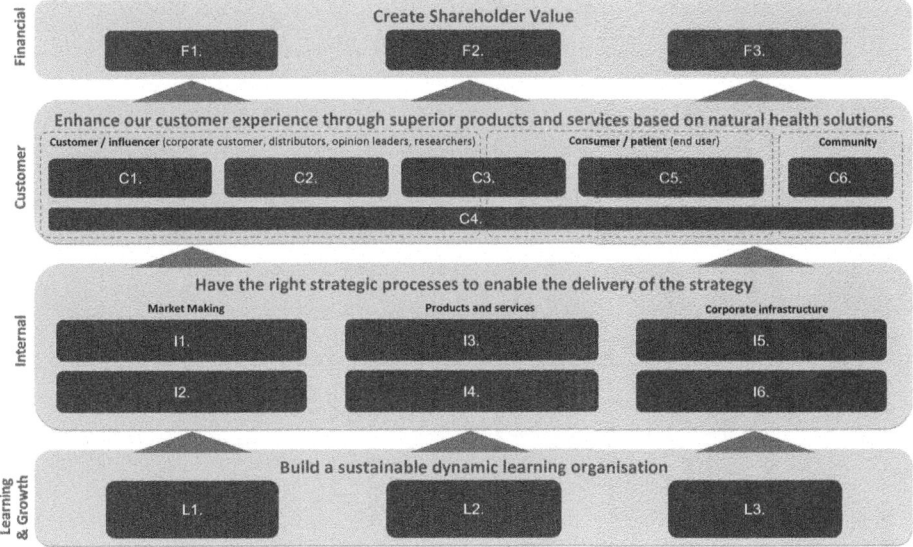

Fig. 1.2 A Strategy Map. (Source: Palladium)

Premium Process (XPP), to align operations with strategy. Figure 1.3 shows the stages of the XPP, with the sub-steps for each stage.

The Balanced Scorecard System, as we shall call it, has evolved continually since its introduction through a seminal Harvard Business Review article in 1992. In an interview with one of the authors, Dr. Norton said, "Bob Kaplan and I launched a revolution, not a static system. It will continue to evolve." That the originators never trademarked the Balanced Scorecard is testament to their commitment to its evolution and the input of others. Over the last couple of years, Dr. Norton has been describing a model for Age 2 Balanced Scorecard Systems (see Panel 1).

Learning from Genetics

As an allegory, think of genetics. There's a strong scientific argument that the human being (or any other being) is not actually that important. It is simply a vehicle for genes to continue to survive and, through genetic mutations, evolve to better deal with the challenges of their changing environments. Strategy management is no different. Whatever the vehicle is, what is important is the thinking that underpins the system. New ideas will be introduced (read genetic mutations). Many will be discarded. Others, which are found to be particularly useful, will be kept. Simple evolution. Stop evolving and you die.

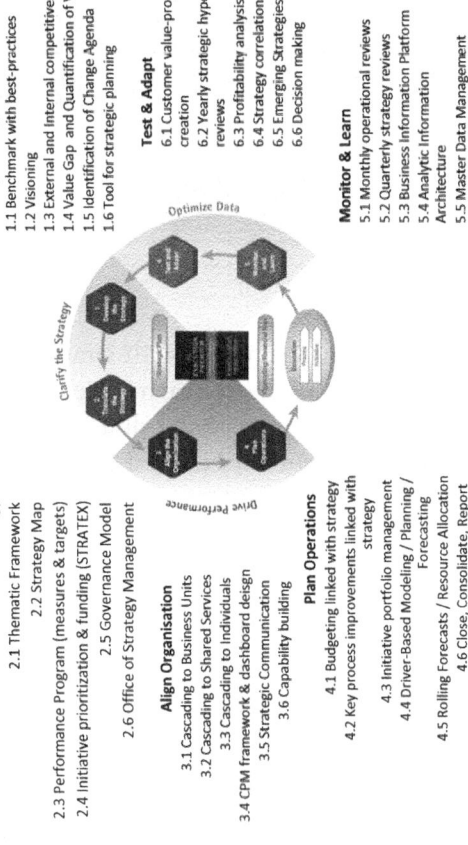

Fig. 1.3 The Execution Premium Process

Based on the authors' many years' experience, research, and hundreds of organizations worked with in shaping and implementing strategic plans, this book adds to the evolutionary, revolutionary, genetics-driven approach. We have myriad observations of successes, failures, and particularly damaging misconceptions. Common lessons emerge, which we will share throughout this book.

Common Challenges

Certain aspects of conventional Balanced Scorecard implementations, as well as the Execution Premium Process, have proven problematic or, more regularly, poorly understood. Kaplan and Norton's last in their seminal canon of five books that traced the evolution of the Balanced Scorecard System was published back in 2008 [2]. As they and others comment, time has moved on, but evolution does not stop. Figure 1.4 shows our *Agile and Adaptive Model for Strategy Management in the Digital Age*.

We have no doubt that systems such as the Balanced Scorecard are still required. The fundamental issues that underpinned the introduction of the scorecard system and other similar frameworks (dealing with more dynamic markets, the potential of technology, etc.) are seismically more challenging now than they were in the 1990s or even the first decade of this century. The

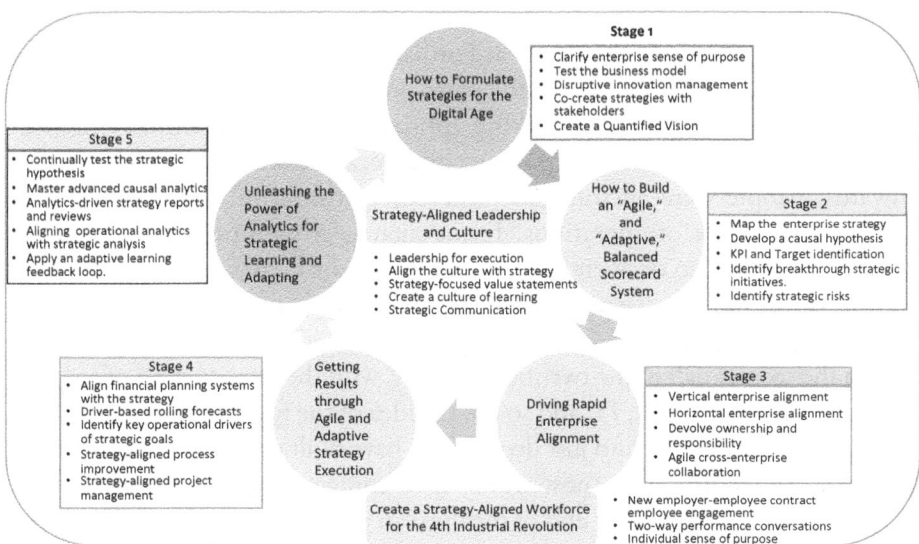

Fig. 1.4 Agile and adaptive strategy execution model

central "story" of the book is how the principles of the Balanced Scorecard System are evolving for these early stages of the digital age. But first, let's step back in time—before the first scorecard article. Indeed, all the way back to 1911. What happened in that year has profound implications for today.

The Scourge of Silo-Based Working

At the turn of the twentieth century, Western economies were facing then-revolutionary change as they began to move from a system of craft-based working to leveraging developments in machinery and automation to enable mass production. Various thinkers grappled with mastering this transition, but the most impactful perhaps was Frederick W. Taylor, because of his 1911 published monograph *The Principles of Scientific Management,* [3].

Faced with increasingly complex industrial models and, at best, a semi-literate workforce, Taylor focused on "simple jobs for simple people." He advocated the training of workers to do very specific jobs and not think outside of that role. He strongly believed that it was the role of management to "think" and workers to "do." More than that, he introduced (or scientifically institutionalized) mistrust as a normal part of management. "Hardly a competent workman can be found who does not devote a considerable amount of time to studying just how slowly he can work and still convince his employer that he is going at a good pace," is how he dismissed the value of an employee and why their "superior" bosses should treat them with suspicion.

Versions of Taylor's system, such as Fordism (introduced by the Ford Motor Company to drive efficiency into the mass-production of Automobiles), institutionalized this "silo"-based working approach and the accompanying fear- and control-based culture. From the second decade of the nineteenth century onwards, people were encouraged (ordered) to think only of their silo—of their narrow sphere of operations. Terms such as cross-functional teamwork, collaboration, and so on were not part of the scientific manager's lexicon.

Such siloed thinking essentially introduced the idea of function-based working. We have since continued to structure organizations according to this template: even in the second half of the twentieth century, with the introduction of strategic planning (see below), HR, IT, and so on, the idea of the function prevailed: this is your job, build expertise, and just do that. We have built professions around these functions. Think also of the now widely used contact centres by banks and the like. The person on the other end of the line can *only* answer questions that relate to their own job (and to a strict script). Nothing outside of that is their job. Therefore, the customer is transferred to another department—and typically must wait on the line for some time: a source of continued irritation to customers.

Frederick W. Taylor would still recognize much of his thinking in today's organizations, despite their espousing of commitment to values, empowerment, and all this nice stuff. Taylor dismissed front-line or factory-floor workers as idiots and not to be trusted. Although appalling, such attitudes were common in the very hierarchical social systems of the time, when only a minority were well educated. However, to be fair, some of the thinking was very useful for transitioning to a systems-based approach to working.

Today, we benefit from highly educated employees that are constantly contributing to and learning from the digital planet (OK, perhaps another cliché). We could ask, "Have workplace attitudes really moved on that much?" Have we stopped encouraging managers to be suspicious of their reports? If so, then why do research firms such as Gallup continually find that only a minority are engaged in most organizations?

Research Findings

According to the 2016 Gallup employee engagement survey, a staggering 67% of employees in the USA are disengaged, a quarter of whom are "actively disengaged" while at work (which, to be fair, is a better figure than many other countries). Actively disengaged means they hate the organizations they work for and will do as much as they can to do as little as they can—Taylor would call this the norm.

Tellingly, the 33% engagement level in the US workforce is the highest it's ever been from Gallup's findings. However, this is only marginally better than the 30% in 2001 (and perhaps no different if we factor in confidence levels and intervals—see Chap. 5: *How to Build an agile and adaptive Balanced Scorecard*, in which we detail common mistakes in working with KPIs).

Such a minor improvement is hardly a ringing endorsement for the billions of dollars ploughed into employee engagement/empowerment initiatives in the interim (and we can assign much of the improvement to improved economic conditions—the 30% figure fell during the so-named credit crunch). Taylorism had a devastatingly negative effect on employee morale, which in the second decade of the twenty-first century is still shockingly low [4].

A Significant Bottleneck

Although perhaps understandable 100 years ago, and being how the Ford Motor Company, among others, delivered their early successes (but even then, the highly divisive and led to employee/company tensions that continue to this day), this functional approach to structuring work is today one of the most significant bottlenecks to successfully managing digital-age organizations.

The functional mind-set encourages (and oftentimes positively rewards) individuals to think only about their specific contribution to the organization's value delivery model—what happens elsewhere "is not my problem." Functions become "personal fiefdoms," which the functional head typically defends aggressively and, of course, fights for the most advantageous share of financial and other resources. It's about promoting the function, not focused on what is required to deliver value to the customer. Look at how budgeting typically works to see the logic of this argument (we explain how the budgeting system must evolve for the digital age in Chap. 7: *Aligning the Financial and Operational Drivers of Strategic Success*).

Seeking Mechanical Solutions

In seeking new frameworks and models, many organizational leaders will still search for something to plug and play—mechanical solutions: again, in keeping with the industrial-age solutions of a century ago. Particularly worrisome here is the continued belief that a software tool can resolve all the issues. Even in these days of advanced data analytics, no single piece of technology will resolve the challenges of strategy formulation and implementation, but can be a powerful aid, when deployed properly, (see Chap. 9: *Unleashing the Power of Analytics for Strategic Learning and Adapting*).

Organizational leaders repeatedly fall for these "instant remedies" that will automatically solve all their challenges, and then scratch their heads wondering why this did not happen.

Strategy management cannot be automated, and no software solution can, or ever will, provide all the answers. Like it or not, managers and staff will also have to "think" and make decisions. Technology can greatly support that decision-making process, but cannot, or rather should not, become that process.

The Strategy Function and Process

So, let's turn our attention to the inherent weaknesses of conventional strategic planning. As cited earlier, when strategic planning was introduced as a new organizational capability, it strictly conformed to the diktats of the mechanical industrial-age mind-set.

As far back as 1994, Professor Henry Mintzberg explained in a Harvard Business Review article that, "the scientific management pioneered by Frederick Taylor …. separated thinking from doing and [created] a new function staffed by specialists, Strategic Planners. The expectation was that planning systems would produce the best strategies as well as step-by-step instructions for carrying out the strategies so that the doers, the managers of the business, could not get them wrong." He went on to say that, "planning has not exactly worked out that way." [5]. Indeed, over the 20+ years since Mintzberg's work, various research projects have found that up to 80% of strategies "fail," often regardless of how well formulated. This, we argue, is due to a fundamental error in how the strategy management process was originally designed and, sadly, is still managed.

Research Evidence

Early in 2016, a research project led by The Leadership Forum Inc., asked more than 200 mid-level managers and their direct reports in a global B2B company a simple question: "What are the three factors that most inhibit the execution of strategy in your business unit or more focused segment of the business?"

What was startling about the replies was not what was said, but what was not. Although many words were applied to the internal barriers to strategy execution (inconsistent leadership, poor communication, lack of clarity, etc.), very little was said about the external barriers. Seemingly, on a day-to-day basis, little thought was given to how marketplace change, especially how the actions of rivals, customers and other actors, could overwhelm their plans.

In the so-named digital age, we speak and write endlessly on the dangers of disruptive technologies, of living in a world characterized by constant change and uncertainty, of small start-ups suddenly and rapaciously eating market share. Yet for all the words, our actions demonstrate that we still cling, with almost religious fervour, to the tried and tested (and typically failed) approaches to strategy execution. That, once we've analysed, and captured in plans, the external world through SWOTs, PESTELs, Five-Forces, and so on, the focus is then solely on getting the inside of the organization aligned to the plan and the units/functions and departments working on delivering their own bits of it. The external world will stand still until the next scheduled planning cycle and resulting strategy refresh.

The fact is that strategic planning and execution are not separate silos in a process—they are part of the same integrated, strategy management process, which also includes learning.

There are many definitions of strategy (see Chap. 2: *From Industrial- to Digital-Age-Based Strategies*), but strategy is essentially about being clear as to what the longer-term strategy destination looks like (a quantified vision helps, *see* Chap. 3: *Agile Strategy Setting*) and, over a period, developing, or more likely enhancing, the capabilities to get there. In today's marketplaces, these capabilities must be operationalized as quickly as possible. But, this is not all. Whatever definition is preferred, an underlying truism that is too often overlooked is that strategy is, as leading strategy thinker Hubert Saint-Onge, Principal of the Toronto-based Saint-Onge Alliance, says, "…a set of assumptions that need to be verified in execution."

Assumptions that Must Be Verified in Execution

Note "a set of assumptions" and "verified in execution." And herein lies a major fault in the strategy management process. Organizations systematically fail to understand that the beautiful plan painstaking crafted over several months must be proven in practice—and that, even in the best cases, reality will uncover flaws in the thinking. With the standard practice being to build the plan and then hand over to managers to implement without deviation or question, the assumptions are untested and the flaws unresolved. As Sir Winston Churchill once said, "However beautiful the strategy we have to sometimes look at the results" [6].

As we explain in Chap. 9, during implementation, advanced data analytics is enabling the testing of these assumptions and, as part of this, understanding the robustness of the hypothesis described in the cause and effect relationships between the objectives and perspectives that populate a Strategy Map. However, organizations must build the capabilities to respond to their findings in a timely manner: not to just capture and discuss at the following year's strategy refresh.

End-to-End Process Management

To overcome this performance-sapping approach, senior management must remove their functional lenses and view strategy management as an integrated end-to-end process.

End-to-end process management drives substantial efficiency and effectiveness gains, perhaps more measurably than any other intervention. This is possibly truer for the strategy management process than any other, given the prize at stake. This is not a new idea, as we have been talking about end-to-end process management for more than 20 years. But, it is not easy to do, due mainly to cultural resistance and fears of losing power and control (the all-important personal fiefdoms). However, when applied, it can really drive breakthrough performance improvements. Of course, the functional work still gets done, but in the context of the outcomes required from the process.

Moreover, end-to-end process management has specific complexities for strategy management. As there will be many strategic thrusts to manage enterprise-wide, strategy execution cannot be viewed as one linear process, but as a collection of related processes. As we explain in Chap. 4, *Strategy Mapping in Disruptive Times*, strategic themes (through which objectives working to deliver the same outcomes) can be powerful aids here.

The challenges of end-to-end process management are becoming even more complex due to how the nature of work is changing. In twenty-first century organizations, increasing amounts of work will be done by "non-traditional" employees (freelancers, contractors, etc.) and organizations will become more boundary-less, in that partners will take up more of the work in delivering increasingly complex value propositions. Within the organization itself, technology will make many processes virtual in nature, with parts done in different geographies and time zones, and held together by digital platforms.

To add to the challenges, end-to-end process management and boundary-less working will necessitate the greater empowerment of knowledge workers—as the virtualized nature of work requires on-the-spot and more real-time decision making—and therefore the further moving away from the rule of the hierarchy and the still too-prevailing idea that management thinks and workers do. What this will look like is still a work in progress.

Somewhat ironically, there will likely be something of a step back to the craft-based working that preceded Taylorism. In the digital age, the crafts will be primarily intangible in nature, as opposed to the tangible, physical work of much earlier centuries. New and rigorous governance models will need to evolve so as to deal with the many complexities of boundary-less working. Dismantling the organizational and people management structures introduced by Taylor and his contemporaries will take some time.

Strategic Innovation

If end-to-end process management is hardly a new concept, it is much more recent than Strategy itself (about 2500 years ago, Sun Tzu said in the "*Art of War,*" that strategy is now too important to ignore, [7]). But even strategy is a new kid on the block compared to innovation.

Humans have been innovating since the big brain developed. Has there ever been a more impactful innovation than the wheel? The ability to make fire perhaps. Yet, innovation is currently a "buzz-word." Today, along with strategy, innovation is "too important to ignore." Consultancies specializing in innovation are flourishing and increasing in numbers (along with that one solution that will make the organization "innovative"). Conferences are selling out.

Disruptive Innovation

When we talk of innovation today, we normally speak of disruptive innovation, which was defined by Harvard Business School Professor Clayton M. Christensen in 1995 as "…innovation that creates a new market and value network and eventually disrupts an existing market and value network, displacing established market leading firms, products and alliances" [8].

However, disruptive innovation did not begin in 1995. For example, as explained in Chap. 2, it was disruptive innovation (and the understanding that organizations sold a "function" or solution and not a specific product or service) that led to the Ford Motor Company not just disrupting but obliteration an industry. As explained earlier, a version of the "scientific principles of management" helped make this happen.

That said, many argue that today, given the breathtaking developments in technology since the mid-1990s, there is no such thing as disruptive innovation. As we also explain in Chap. 2, if you are being disrupted, you are simply *not paying attention.* This anchors back to organizations needing to pay more attention to external changes as they execute a strategy and seeing strategy as not just a single "end-to-end process," but as dynamic, in which various stages feed backwards and forwards into others. Experienced strategy execution practitioner, Mihai Ionescu, Senior Strategy Consultant at the Romania-based Strategic Systems Consulting (Strategsys), and one of a select few advisors/ practitioners interviewed for this book, comments:

So far, strategy management has been regarded more as a chain of discrete, sequential processes, while the operational management is essentially seen as a continuous process. It might not be like this anymore. The 4th industrial revolution is steaming ahead, enabling us to handle time, distances and information like never before, and allowing us to validate hypothesis about the future with more accuracy and much faster. A continuously deepening VUCA [Volatility, Uncertainty, Complexity and Ambiguity] requires strategy management techniques that don't really exist today, because those currently in use have been created on the waves of the past industrial revolution.

He adds that. "Instant communication, collaboration and co-creation, plus the anywhere-anytime access to rich, diverse and loosely-correlated information, can fundamentally transform the sequential formulation-planning-execution annual cycle into a continuous process, where all stages of the Execution Premium Process overlap and interact in time, as some inter-dependent swimming-lanes that exchange information and decisions along the way."

The Importance of Agility

As much as anything, Ionescu's observation speaks to two performance dimensions (that, as with most requirements today, meld into one) that are critical to succeeding in these early days of the 4th industrial revolution, in addition to the need to strategy as a single, inter-dependent process: agility and synchronizing the internal rate of change with the external rate.

An agile enterprise can innovate, drive transformation change, and be flexible, whilst *also* maintaining a strong focus on strategy and on the customer.

A useful definition of agility comes from the US-headquartered benchmarking firm, The Hackett Group, "[agility is] the ability of an organization to synchronize the internal rate of change of the business with the rate of change imposed by the external business environment."

> An agile enterprise synchronizes with the external rate of change by inculcating, and constantly adapting, scalable and customer centric operations and digitally connected value chains that cut through hierarchies and organizational silos. They also leverage data analytics to capture and transform data into actionable knowledge that drives proactive decision making, [9]

An Agile and Adaptive Model for Strategy Execution in the Digital Age

This brings us to outlining the five stages (and a central steer and underpinning requirement) of our *Agile and Adaptive Model for Strategy Management in the Digital Age*, which we describe fully in the following chapters.

Agile and Adaptive

Agile points to sudden quick changes—being "able to move quickly and easily" according to the Cambridge English Dictionary, whereas it defined adaptive as "having the ability to change to meet different circumstances" (which does not necessarily mean quickly or easily).

Stage 1, How to Formulate Strategies for the Digital Age

Chapters 2 and 3 consider how the process of strategy formulation and planning is evolving for the digital age. It needs to be much quicker than has generally been the case historically, and with a lighter touch, while also better involving key managers that must implement the strategy, as well as planners.

Moreover, there is a requirement to define the organization's "sense of purpose." This should be captured within a mission statement (and which, unlike vision statements, are not time bound and rarely change).

Furthermore, there is a need to capture the voices of the customer, as well as other key stakeholders, so that all involved in the value chain are on the same page. Tools such as customer co-creation, in which customers are actively involved in shaping solutions, play an important role here—as well as stakeholder analysis. We will also explain how other newer approaches such as blue ocean strategy, business model innovation, technology-based planning, situational analysis, the OODA Loop, and so on, are helping create more relevant and agile/adaptive strategic plans.

Also explained will be the importance of developing both longer-term and medium-term quantified visions that include (1) a quantified success indicator (a global benchmark perhaps, or revenue target), (2) a definition of the niche (where to compete), and (3) a designated timeframe (within two or five years). We also stress the value of creating a Strategic Change Agenda that describes the critical performance dimensions that the organization must master to deliver on the strategy with defined current, to desired, states.

Stage 2, How to Build an Agile and Adaptive Balanced Scorecard System

Chapters 4 and 5 take us into the creating of the Balanced Scorecard System, which comprises a Strategy Map and scorecard of Key Performance Indicators (KPIs), targets and initiatives. Although the original structure, as described by Doctors Kaplan and Norton, remains largely intact, we describe a process for more rapidly building scorecard systems—with less of a focus on building a "perfect" scorecard system that many organizations expect to simply plug and play. We emphasize the role of analytics to provide more of a useful steer to agile and adaptive execution and for continued testing of the assumptions of the causal relationships with the system.

Furthermore, we explore many of the challenges that have often stymied attempts to build optimal scorecard systems, such as the absence of robust objective statements and not using tools such as driver-based models and Key Performance Questions to bridge the gap between objectives and KPIs. Another area that has been problematic for organizations is a general lack of understanding of the basics of the science of measurement, leading to their making, sometimes expensive, decisions based on what *is believed* the data is saying rather than what *is actually being* said.

Stage 3, Driving Rapid Enterprise Alignment

Historically, this has oftentimes been a slow and cumbersome process, top-down and imposed. This is increasingly problematic in today's fast-moving markets. By the time the process is complete, at least some parts of the strategy are often out of date, or some of the assumptions on the corporate level Strategy Map are proven to be false, or at least only partly true. Turning around this beast of an enterprise scorecard system proves extremely challenging and performance sub-optimizing (and often not attempted until the next strategic planning cycle).

Moreover, experience has shown that the conventional approach has often led to resistance from staff, who see it as little more than another measurement control system that they wished would disappear.

In Chap. 6, we outline an approach for identifying the critical (and very few) objectives and KPIs to devolve (the spine of the organization) and then empowering teams to build their own scorecard systems that describe what they want to achieve over the coming period. As well as leading to greater buy-in and ownership, this transmits a message that senior management trusts their employees and believes in their abilities (Taylor must be spinning in his grave!). Proper governance still ensures alignment, but guided by flexibility and empowerment instead of rigid imposition.

Stage 4, Getting Results Through Agile and Adaptive Strategy Execution

In and of themselves, Strategy Maps and scorecards do not execute strategy, but are, rather frameworks for articulating the objectives, identifying KPIs, and so on. Indeed, the scorecard system is the final part of strategic planning. In Chaps. 7 and 8, for strategic plans to be implemented successfully, they must be effectively linked to financial planning, as well as strong project and process management capabilities.

As strategic planning is no longer fit-for-purpose for the digital age, this is equally true for conventional financial planning processes, particularly the annual budget, which is time-consuming, overly detailed, and generally locks funding for a calendar year. Agility is missing. As a result, by the time they are published budgets are typically out-of-date and set funding makes it a challenge to quickly allocate/reallocate resources in response to emerging external opportunities and threats. What is required is a shift from the conventional budgeting process to a system based on driver-based rolling forecasts, which enable the agile allocation of financial resources. Moreover, from a strategy viewpoint, we recommend hardwiring this to a mid-term plan and work in tandem with a Strategy Map and Balanced Scorecard.

Similarly, organizations need to be better at prioritizing the process improvement activities according to strategic needs, using models such as driver-based models to provide a more precise link between strategic goals and operational improvements.

Stage 5, Unleashing the Power of Analytics for Strategic Learning and Adapting

The next evolution of the Balanced Scorecard system is from being primarily a communication/alignment model to a framework for powerful performance analytics. As we explain in Chap. 9, advanced analytic tools are enabling more precise testing of the causal assumptions embedded within a Strategy Map, enabling both descriptive (what has happened) and predictive (what is likely to happen) analysis. Doing so also enables agile allocating of resources to the process improvements and other interventions with the greatest impact on desired results.

Furthermore, such analytics enables organizations to identify best (and poor) practices and quickly share these across the enterprise. In the digital age, knowledge management systems that are truly strategy-focused and analytics-based will become a key competitive differentiator.

At the Centre of the Model: How to Ensure a Strategy-Aligned Leadership and Culture

At the centre of an Agile Strategy Execution Model is leadership and culture. Put simply, leadership and culture are indivisible and, if these are not synchronized and strategy-focused, then strategy execution will flounder and likely fail. In Chaps. 10 and 11, we describe the components of "leadership for strategy execution" as well what a "strategy-aligned" culture might look like. We explain how to shape meaningful value statements (rather than nice sounding words that hang on walls and are generally ignored) and the importance of ensuring structures, policies, processes, decision rights, and information flows support the desired values. Cultural assessment tools, that leverage advanced data analytics capabilities, enable a more precise understanding of cultural gaps and bottlenecks.

Communication is an important aspect of getting the culture right. It is a key management discipline in any circumstance, and especially critical when an organization is setting out to implement strategy. We stress, however, that communication should be an ongoing process, rather than a one-off exercise repeated on an ad hoc basis. Messaging must be a constant part of reinforcing the dos and don'ts around strategy. If this is not done, there is a pressing danger that decision makers, and indeed all employees, might revert to inappropriate behaviours. The mantra "communicate, communicate, communicate" is commonly heard, but less often acted upon.

Underpinning the Model: Creating a Strategy-Aligned Workforce for the 4th Industrial Revolution

Knowledgeable employees are probably the key to organizational success in the 4th industrial revolution. In Chap. 12, we explain that there is a requirement for a fundamental, indeed transformative, shift in the very essence of how we view the employee-employer relationship. Firms are increasingly confronted with a generational mind-set shift that values learning and engagement much more than job security. Millennials, for instance, are looking for a very different workplace experience than earlier generations. Indeed, they are looking for a "sense of purpose" focusing on how their own self-esteem and worth is heightened through the work experience, as well as how their learning and growth (and therefore market value) is engendered. In addition, we have little sense yet of the expectations of a post-Millennial Generation that have spent their whole lives living in a fully connected, digital world: for this generation, connectivity is their "mother tongue."

Organizations need to inculcate mechanisms for aligning the individual's purpose with that of the organization's, as captured in the mission. This is the essence of engagement.

To achieve this, organizations must rethink the traditional job contract, focusing more on aligning the individual's "sense of purpose" with that of the enterprise (and for many employees, only for a relatively short timescale); greater value will be extracted when we think of how employees and organizations *work together*, as opposed to employees *working for* organization—a digital-age concept perhaps, but in many ways little different from the craft-based structures of the yesteryears.

Moreover, we explain processes by which organizations can ensure that they have, and are developing, the skillsets and competencies required for delivering to the strategy, such as through strategic human capital readiness frameworks.

Collaborative Scorecards

In Chap. 13, we explain that with cross-organizational partnerships becoming increasingly prevalent in driving customer value, how Strategy Maps and Balanced Scorecards are being developed to align various organizations (from focused partnerships to mergers and acquisitions) along common goals and themes and that cut across an organization's functions, departments, and so on. Also explained is how technology is powering collaborative working through the scorecard system: used extensively to drive "social value" and transform the lives of communities in emerging economies and deprived regions.

Parting Words: Shifting Paradigms

To end the chapter with a further cliché, we are in an age of "shifting paradigms." How we manage and structure organizations and how we go to market are, and will be increasingly, fundamentally rethought for the digital age. Yet, strategy management is more important than it ever has been throughout its 2500+ year history. Organizations always need goals and must make the right choices for their achievement. This is the essence of strategy. However, organizations need to better understand that strategy is a set of assumptions that must be verified in execution—for this, the fundamentals of the strategy management process must be reconfigured and away from the conventional "planners plan and managers execute" mind-set.

Finally, strategy management can no longer be seen as an engineered, sequential process, but as a dynamic system in which all parts continually feed into each other. In short, strategy management, as we reinforce throughout this book, must become agile and adaptive (and know when each is applicable) with the appropriate governance model to ensure that execution of plans can be modified as required without leading to chaos. As General (later President) Dwight Eisenhower once said, "The plan is nothing, planning is everything" [10].

> **Panel 1: Dr. Norton's Age 2 Balanced Scorecard System**
>
> **Age 2 Balanced Scorecard Systems**
>
> Over the last couple of years, Dr. Norton has been evolving a model for Age 2 Balanced Scorecard Systems, which adapt the system for the opportunities and challenges of the digital age (identified by an analysis of mega-trends affecting global economic development and working). As shown in Fig. 1.5, dealing with these mega-trends requires organizations to evolve from being strategy-focused to strategy-learning organizations, which focuses on five dimensions.
>
> 1. **Networked:** Aligning networked organizations in a world that is increasingly boundary-less. Networks are both within the organization (and across functions and regions) as well as between organizations. Collaborative technology, supported by good governance and more empowered working, is critical to ensuring such networks are effective.
> 2. **Human Capital:** Developing human capital capabilities in world economies increasingly based on the capabilities of employees (and that are part of the networks).
> 3. **Shared Value:** In a world filled with shared value, using the scorecard system to build partnerships to drive benefits for the commercial organization as well as society and communities (becoming good corporate citizens).

4. **Risk:** In a world filled with risk, aligning risk with performance through creating risk dashboards that track the risks that impact each objective on the Strategy Map.
5. **Analytics:** In a data-rich analytic world, using the power of advanced data analytics to test the causal assumptions that underpin the strategy and as captured in the strategic objectives, KPIs, and initiatives.

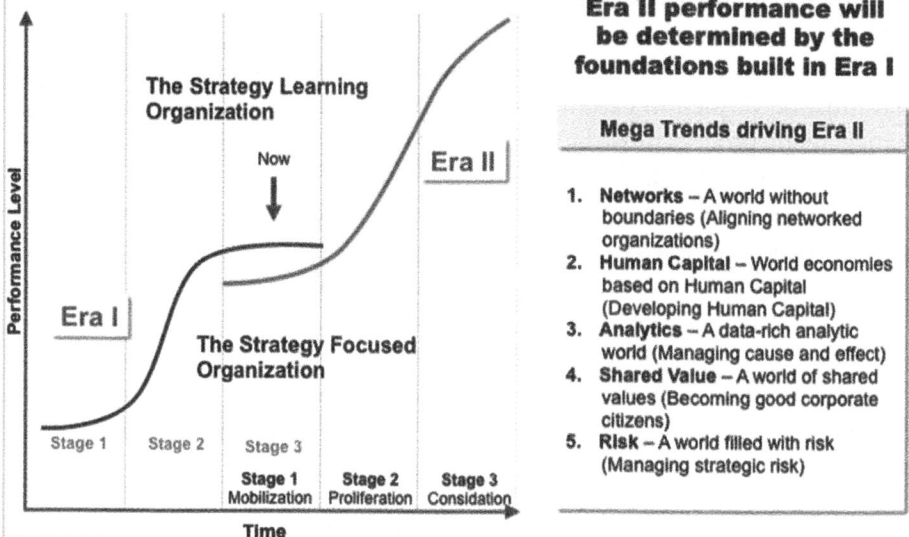

Fig. 1.5 Age 2 Strategy Management System

We discuss each of these Age 2 components in subsequent chapters. But note, Norton comments that Age 1, which was essentially about building scorecards, communicating strategy, and alignment, are still important and the foundation for Age 2. It's about evolution, not obliteration.

Self-Assessment Checklist

The following self-assessment will assist the reader in identifying strengths and opportunities for improvement against the key performance dimension that we consider critical for succeeding with strategy management in the digital age.

For each question, any degree of agreement to the statement closer to one represents a significant opportunity for improvement (Table 1.1).

Table 1.1 Self-assessment checklist

Please tick the number that is the closest to the statement with which you agree		
	7 6 5 4 3 2 1	
In my organization, strategic planning and execution are part of the same integrated strategy management process		In my organization, strategic planning and execution are separate silos in a process
My organization shows a high level of agility and adaptiveness in the strategy management process		My organization shows a low level of agility and adaptiveness in the strategy management process
The senior leadership team fully understands that "strategy is a set of assumptions that must be verified in action"		The senior leadership team poorly understands that "strategy is a set of assumptions that must be verified in action"
During strategy execution, we continually survey and respond to external changes		During strategy execution, we rarely survey and respond to external changes
My organization has an established strategy execution framework, such as a Balanced Scorecard or similar		My organization does not have a strategy execution framework such as a Balanced Scorecard or similar

References

1. Robert Kaplan and David Norton, *Measures that Drive Performance*, Harvard Business Review, January/February 1992.
2. Robert Kaplan and David Norton, *The Execution Premium: Linking Strategy to Operations for Competitive Advantage*, Harvard Business School Press, 2008.
3. Frederick W. Taylor, *The Principles of Scientific Management*, Harper and Brothers, 1911.
4. *Employee Engagement Survey*, Gallup, 2016.
5. Henry Mintzberg, *The Fall and Rise of Strategic Planning*, Harvard Business Review, January/February 1994.
6. Sir Winston Churchill, *Attrib*.
7. See Sun Tzu, *The Art of War, Special Edition*, translated and annotated by Lionel Giles, El Paso Norte Press, 2005.
8. Joseph L. Bower, Clayton M. Christensen, *Disruptive Innovation: Catching the Waves*, Harvard Business Review, January-February 1995.
9. See Roy Barden and Elizabeth Watts, *Public Sector Innovation Summit 2017 Report*, The Hackett Group, May 2017.
10. Dwight Eisenhower, speaking to the *National Defense Executive Reserve Conference* in Washington D.C. on November 14, 1957.

2

From Industrial- to Digital-Age-Based Strategies

Introduction

In the next two chapters, we describe how the process of strategy setting is evolving for the digital age. This chapter considers what is meant by strategy, the criticality of understanding the longer-term "sense of purpose," understanding the "function," that the organization delivers to customers, business model innovation (and why disruptive innovation is something of a myth), and the importance of bringing stakeholders (most importantly customers) into the strategy formulation process. The next chapter looks at setting shorter-term strategic destinations; including shaping quantified short- and longer-term vision and identifying the value gap (present and desired states) (Fig. 2.1).

The Potential Dangers of Agility

The urgent requirement for organizations to be more agile in their strategy management process is a central message of this book and of the model that we propose. That said, we must be somewhat cautious in religiously applying agile thinking to each and every step of the process.

To explain, agile thinking reaches back many decades and has its roots in the lean/total quality thinking of the 1980s. However, it gained significant traction and general acceptance through the publication of the *Agile Manifesto* in 2001, which sought to deal with the then rising frustration with failed software development projects.

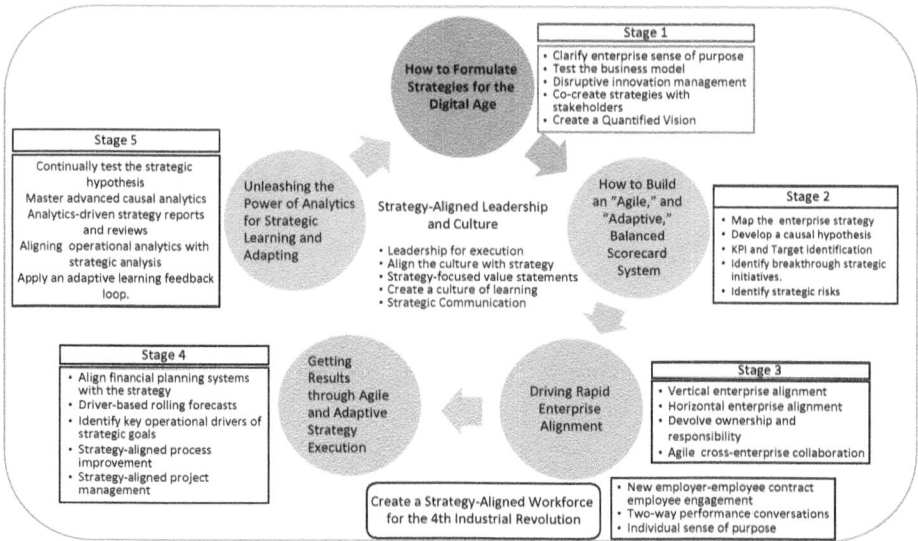

Fig. 2.1 Stage 1: How to formulate strategies for the digital age

Now, much of what emerged from the thinking that went into the Agile Manifesto is sensible. We certainly concur with this comment from two of the key originators, Martin Fowler and Jim Highsmith, that, "No one can argue that following a plan is a good idea—right? Well, yes and no. In the turbulent world of business and technology, scrupulously following a plan can have dire consequences, even if it is executed faithfully. However carefully a plan is crafted, it becomes dangerous if it blinds you to change" [1].

This argument is even more relevant in the second decade of the twenty-first century than it was at the start of this first. The same can be said of the essential message of the Manifesto's four values (such as customer collaboration over contract negotiations) and 12 principles (such as collaboration between the business stakeholders and developers throughout the project).

Be Careful with the "Sprint"

Central to the agile methodology (and as practiced extensively in project management circles) delivery is typically through incremental, iterative work sequences that are commonly known as sprints. Since first introduced to software development, agile thinking and the seductive image of the "sprint" have been somewhat ubiquitously applied to organizational processes or activities—agile innovation, as one example. It is also being applied to strategy management—this is where we must take care.

James Coffey, Principal of the US-based Beyond Scorecard (and previously a colleague of the authors) makes this useful observation: "Strategy is not like coding. Agile software development involves rapid coding, testing and rework to develop in an iterative manner, relying on rapid feedback throughout the process."

He continues that this is fine in a closed environment where it is straightforward to design and test against requirements, however, "Strategy exists in an open, dynamic environment where outcomes are often impacted by exogenous events where definitive outcomes may not be known for an extended period. You cannot break strategy into small chunks and constantly reassess and change the plan, since each piece of the strategy is interconnected in a variety of ways; changing one will have unforeseen consequences on the rest."

> As a result, strategy evolves and changes in a different manner, with the whole strategy considered and changed based on the environment and the impact the change has on the entire strategy.

Coffey concludes with, "Agile answers the question, 'What do we need to do and change to match this set of demands?' while strategy answers the question, 'Is what we are doing still relevant in the current environment?' In Agile, you can define requirements and the development adapts to them; with strategy the environment sets the requirements and you must adapt to them."

The fact is, we need to be both agile and adaptive and understand the difference. Agile points to sudden quick changes—being "able to move quickly and easily," according to the Cambridge English Dictionary, whereas it defined being adaptive is about "having the ability to change to meet different circumstances" (which does not necessarily mean quickly or easily).

Challenging the Notion that "Strategy is dead!"

To a large extent, agile thinking (when considered against the background of fast-moving, unpredictable marketplaces) has led to the rising tide of commentators promulgating that "strategy is dead" or more specifically, that the shaping of long-term plans is no longer fit-for-purpose. Many alternatives are about focusing on short-term plans only (a version of the sprint). Now, there is a kernel of something useful here as we explain in subsequent chapters. But first, let's debunk the death myth.

It is bemusing that the validity of strategic planning has been questioned because we compete and operate in a complex, unpredictable, and fast-moving global economy. Strategic planning as a discipline emerged in the 1960s and early 1970s *precisely because the world was becoming less predictable*. Most notable was the experience of Royal Dutch Shell, where the deployment of strategic planning techniques led to their predicting the formation of The Organization of the Petroleum Exporting Countries (OPEC). Consequently, Royal Dutch Shell gained considerable market share during the oil crisis of 1973. Other companies promptly took note. Unfortunately, as we argued in Chap. 1, shaping strategic planning according to the functional diktats of F.W. Taylor was a mistake.

Yet, compare the unpredictability of the world in the early 1970s with that of today. Unpredictability has become the norm. As a result, we argue firmly that strategy and strategic planning are required more than ever.

Defining Strategy

Now, here, we must provide some clarity around definitions. An oddity of the strategy world is that there has never been a universally agreed definition of the term *strategy*. Therefore, people can argue that "strategy is dead," but it is rarely clear what exactly they are consigning to the graveyard. Is strategy an end-place? A group of choices and plans to get to that destination? Or a combination of these—or something else?

From our experience, this in and of itself is a potential bulwark to successful strategy management and execution. If the senior management team has not agreed what strategy is, how can it be managed or implemented?

Case Illustration: A Singapore Clothing Manufacturer

A few years ago, one of the authors interviewed a CEO of a clothing manufacturer in Singapore that had succeeded with a Balanced Scorecard program (i.e. delivered to the strategic goals) and asked, "If you were to start again, what would you do differently?" He replied, "That's easy, I'd get the leadership team to agree what we collectively meant by the term strategy and write it down. I really struggled to lead the crafting of the strategic plan until I realized (in a flash of inspiration while watching soccer at home) that we were all talking about completely different things."

He provided this illustration, "On one side of the table I have a manufacturing director, whose view of ideal (and so his understanding of what strategy should be about) is two colours of jeans in two sizes. Easy to plan and manage. On the other side of the table, I have a marketing director who has to sell these jeans to teenage girls who change their mind about their size and preferred colours every 15 minutes. Very difficult to plan and manage. These two directors live in different worlds, yet both had valid observations and crucial contributions to make. So, I had to get them and the rest of the team to agree on what strategy meant and how it must be delivered."

Harvard Business School Professors: Useful Definitions

So, what does strategy mean? There are many candidates, some more useful than others. The esteemed Harvard Business School Professor Michael Porter defines strategy (or more precisely, competitive strategy) as about being different. "It means deliberately choosing a different set of activities to deliver a unique mix of value," he has said [2].

Porter argues that strategy is about assuming a competitive position, about differentiating yourself in the eyes of the customer, about adding value through a mix of activities different from those used by competitors. To Porter, the essence of strategy "is choosing what not to do" [3]. Porter has said that if a leadership team cannot answer the question, "what don't you do" in their industry, then they do not have a strategy.

Porter's Harvard Business School colleague Professor Jan W. Rifkin has defined strategy as, "an integrated set of choices that position a firm, in an industry, to earn superior returns over the long run." Let's break this down.

An Integrated Set of Choices

Strategy is about *choice*: as Porter said, this is both what to do and, importantly, what not to do. Selected choices should work together (integrated) in delivering value and sustainable competitive advantage.

This highlights one major weakness of the strategy "sprint." Focusing on the short term only will likely lead to not making sensible choices and the continuous grabbing at disparate opportunities (go—or sprint to—where the money is NOW) and so not being disciplined in what not to do—resulting in an inability to properly develop and nurture a sustainable differentiator.

Position a Firm, in an Industry

Knowing precisely what your value differentiator and capabilities are enables the expending of energies in the right places. Expertise and capabilities are further developed, and reputations enhanced—a good place in which to be.

Over the Long Run

Organizations do not exist just to deliver the short-term numbers (although the lunacy of the quarterly reporting hysteria would suggest otherwise). Where you want to be tomorrow is as important as where you are today. The trick is accelerating the speed of how you get to the desired destination and the agility/adaptiveness to switch paths on the way to that envisioned place.

> **Advice Snippet**
>
> A useful first step in any strategy formulation exercise is to get the senior team to agree on what they mean by the term *strategy*. Without agreement, what possible chance do they have of shaping the most appropriate objectives or initiatives? And how can buy-in possibly be achieved? We recommend the following steps:
>
> 1. Each individual leader writes down on a piece of paper what they understand by the term strategy.
> 2. Each participant then reads out their definition (the differences can be astounding although common themes will be evident).
> 3. The team discuss and agree upon what strategy means for the organization.
> 4. Capture the agreed statement and use it in each strategy review session and for communication purposes.

Defining the Sense of Purpose

A fundamental question for all organizations to answer is, "why do we exist?" Many would casually dismiss this question with, "to make money and deliver shareholder value," (if a commercial organization) and, for any enterprise, "to deliver a product or service to a customer." Well, yes, but that's not the whole story.

The Mission Statement

Let's start with the mission statement. This defines the purpose of the organization: why it exists. A mission statement generally doesn't (or at least shouldn't) change much over time, and cannot possibly be agile.

Google's mission "To organize the world's information and make it universally accessible and useful" has been stable since the firm's launch in 1998:

this in an industry that is at the forefront of rapid change and "disruptive innovation." Note too that the mission makes no mention of the internet or of search engines, so developing self-drive cars does not conflict with the mission.

Google has always understood that it does not exist to offer a narrow product or delivery vehicle, although it recognizes the requirement of being adaptive within the confines of the mission. We argue that neither the product nor the delivery vehicles should be included in such a statement.

Case Illustration: The Ford Motor Company (Disruptive Innovation in the Early Nineteenth Century)

To illustrate this, we go back in time to the beginning of the twentieth century. Henry Ford famously said, "If I had given customers what they asked for, I would have given them faster horses" [4]. He realized that people weren't actually buying horses and carriages, but a "function": that is, the function of carrying people safely and quickly to their destination. Automobiles (that became known as *cars* due to people's difficulty in moving away from the term *carriage*) did that much better. Ford saw this, and the mass production of cars began.

The automobile essentially put carriage makers as well as the providers of horses out of business—today, we would call this "disruptive innovation."

Case Illustration 1: Kodak (Not Paying Attention to the Function)

A modern-day equivalent is Kodak, who stupidly held on to the fictitious belief that their customers wanted to buy film or film processing. They did not; they were simply available vehicles for meeting a function—capturing memories, and so on. Moreover, when Kodak started to suffer the consequences of digital technology (which ironically, Kodak essentially invented) they began to enable customers to capture photographs on a DVD and then send them out for processing, so clinging onto a key strand of their belief system and their treasured century-old business model.

In his book, "*The Decision Loom: A Design for Interactive Decision-Making in Organizations,*" Vince Barabba, a former Kodak executive, offered insight on the choices that set Kodak on the path to bankruptcy. One extraordinary quote comes from Steve Sasson, the Kodak engineer who invented the first digital camera in 1975. Management's reaction was apparently "that's cute – but don't tell anyone about it" [5]. It would take several decades more before Kodak would fully understand that "don't tell anyone about it" is not a sensible long-term strategy. They had over 25 years to prepare for the cataclysmic disruption that they apparently "didn't see coming!" or more honestly—purposefully chose to ignore, in the hope, we guess, that no one would notice.

Case Illustration 2: Blockbuster (Not Paying Attention to the Function)

A further example is Blockbuster, the once dominant supplier of rented videos/DVDs who were put out of business by Netflix (who Blockbuster could have purchased in 2000 for $50 million and which, as of May 24, 2018 had a market capitalization of $153 billion).

Blockbuster foolishly believed that customers *wanted* to go to stores and line up to rent a film. Moreover, they were hit with fines if they returned the rentals late, which was a significant revenue stream for Blockbuster—we would suggest that relying on extensive income by punishing customers is not a good place to be. Netflix offered the primary function (enabling customers to view films conveniently and at their leisure) through online streaming and subscription (so no late charges). The Blockbuster horse and carriage was out of business.

There are many other recent big-name examples of this disastrous belief that it is all about the product and a refusal to change the underlying business model. Unquestionably, there will be many more to come.

The Relevance Question

Therefore, perhaps *the* key strategic question that senior executives must constantly ask is what we call the "relevance" question: *Is what I am making or selling still relevant, and will it be relevant five or ten years from now?* Relevance means constantly questioning whether how the function is being delivered is under threat as well as the strategy and its execution hypothesis. Adaptiveness is required—which is essentially what evolution is (evolution is never agile).

> **Advice Snippet**
>
> To begin to understand that the organization exists to deliver a function and not a product or service, we ask leaders to answer two questions and stipulate that they require different answers.
>
> 1. What products and services do you sell?
> 2. What do you sell?
>
> While the first question elicits many answers, the second is usually met with blank expressions (as people typically think it's the same as question 1). Oftentimes, getting to the root of the function is a difficult and painful conversation for leaders—one CEO of a large telecoms firm says that their 9/11 was the day Apple released the iPhone, as they realized that they would not have known how to do this (or even why it was important). They had failed to realize that a cell phone was not simply a mechanism for making telephone calls, but a personal communication system. Apple got this—as did Ford and Netflix in other domains.
>
> So, always keep front-of-mind that the products and services the organization sell is *not* the same as what they sell.

Reinventing the Business Model

Although being cognizant of the disruptors that are shifting the market conditions (competition, distribution, regulation, customer behaviours, macroeconomic factors, etc.) is critical, it is just the beginning of shaping a robust strategy. What's next is for leaders to ensure the business models are fit-for-purpose.

Hanging on to outdated business models is hardly new, but the impact of doing so is much greater than ever before. The cited Kodak and Blockbuster experiences are examples of how these drivers play out: total refusal to abandon a strategy and business model that had worked very well (in Kodak's case, for over a century); failure to monitor (or accept as important) the changes in customer behaviour; and a veritably ostrich-like "head in the sand" response to a dramatically changing world. The brand goes stale, core strengths languish, and opportunities to keep pace with—or get in front of—changing customer needs pass them by. The world evolves; they don't.

Research Evidence: A Gloomy Picture

2014 research by The Palladium Group (for whom the authors of this book were then working) painted a gloomy picture of how organizations were dealing with significant change. Fully 82% of the near-3000 organizations surveyed (and from across the globe) agreed that changes in their business environment were intense, and 72% believe their business model will be under threat in the next five years (61% state that this was already the case).

Worryingly, although leaders knew that the danger comes from outside the organization, few were doing much about it, with only 24% of organizations constantly monitoring their competitive environment. Tellingly, those that did were 6.7 times more likely to achieve breakthrough financial performance than those that did not. Furthermore, only 14% regularly updated their strategies in response to changes in the market.

Although most research respondents (93%) agreed that innovation is a key factor to their future success, just 36% believed they were good at innovating. Indeed, most respondents believed they were poor at all forms of innovation studied, including incremental innovation (product/service and process innovation), customer experience innovation, and business model innovation (reinventing aspects of an organization's value proposition or operating model). Sixteen per cent did not use any innovation at all. In a time of ongoing disruption, these numbers are concerning, to say the least.

Disruptive Innovation: Something of a Myth

Developing strong and innovative strategies requires a mind-set shift from an internal bias to an external bias. We must bias our decision-making process towards what the customer/market/environment is telling us, so to understand which key trends and advances will most affect the value delivered. Yet, much of what we see in the current batch of innovation offerings does not do this well. Actually, a lot of nonsense is promulgated about innovation.

Just about every consultancy has an offering that promises a "proven" approach (the much sought-after plug and play solution) to becoming a master innovator and consequently reaping rich rewards. New models, ideas, and some frankly bizarre measurement systems are explained in books and are "wowing" delegates at conferences and workshops. Self-proclaimed "Thought Leaders" or gurus are building empires. Conferences are packed to the rafters. It's big business and will only get bigger as there's a big demand.

We would argue, however, that there is no such thing as disruptive innovation. If you are being disrupted, you are simply *not* paying attention. So, here's a recommendation that many, at first, might find odd. We recommend that organizations should forget business model innovation, process innovation, product innovation, and so on—at least as starting points.

True innovation does *not* start from innovation. It starts from competitive advantage and understanding, through powerful mapping systems that already exist, the technologies that drive this (and not just advanced digital technologies), gaining clear insights into how technologies are being deployed and combined elsewhere (and don't be myopic and focus on your own industry), and plan from this position. The goal should be to constantly outmanoeuvre the competition—that's competitive advantage.

Technology-Based Planning: Outmanoeuvring the Competition

In a LinkedIn blog in 2016, consultant Iain Wicking provided powerful insights into this from the viewpoint of technology-based planning. "Via technology-based planning, a "disruptive technology" is no more than an observable manoeuvre in "technologyspace" that can be easily countered in a number of ways. Given this, disruptive technology innovations are an illusion and the myth peddled by so-called thought leaders that these innovations cannot be dealt with proactively.

We should be able to see disruptive influences emerging well in advance and plan defensive or offensive manoeuvres accordingly" [6].

New Wine in Old Bottles

Technology-based planning might seem like a digital age concept and something only of interest (and understandable) to technological whiz kids. Yet, and once more referring to the multiple influences of the early twentieth century Ford disruption success story, this is precisely the approach they took (without the sophisticated digital mapping capabilities of course).

Moreover, (and with very different technologies than Ford) today, a handful of some of the most successful companies, such as Apple and Microsoft, were built on a form of technology-based planning. Steve Jobs intuitively understood that the foundation for building, growing and maintaining an organization had to be technology acquisition and utilization. He artfully manoeuvred in the technologyspace and did it mostly in his head. For this, he was considered a genius. Giving that technology-based planning will be a new (and even alien) concept to many readers steeped in the MBA-style worship of finance-based planning, we have outlined what it looks like in Panel 1.

Finance-Based Planning

Secondly, and strongly linked to the first point, most (just about all) of the work we see in innovation is by planting it firmly within the confines of established finance-based planning models. Innovation becomes a standalone program, a number of lines on a spreadsheet, something to debate in internal sessions and then the findings retrofitted to the Tayloresque business models so loved by many. And then we build ridiculous measurement systems to assess innovation and hold brainstorming sessions to "innovate"—then, oops, we get disrupted and wonder what happened!

As we continually argue, as we sequence through the early years of the digital economy (or the 4th Industrial Revolution, or whatever name future historians give to it), we must realize that competitive advantage will not be sustained simply by making industrial age models work better. This is not sustainable. No ideas, frameworks, or words of wisdom from today's "Gurus" that still root themselves to the spreadsheet-driven tyranny of economic/finance-based models will fix this. As Professor Albert Einstein famously said, "We cannot solve our problems with the same thinking we used when we created them."

But let's consider some disruptive approaches that are better known and more generally applied than technology-based planning.

Blue Ocean Strategy

In 2004, W. Chan Kin and Renee Mauborgne introduced the concept of Blue Ocean Strategy. In their book *Blue Ocean Strategy: How to Create Uncontested Market Space and Make the Competition Irrelevant*, the authors differentiated between competitive spaces known as red oceans and blue oceans. Red oceans represent all the industries in existence today or in the known market space. "In red oceans industry boundaries and defined and accepted. Here companies try to outperform their rivals in order to grab a greater share of demand."

Blue oceans denote all the industries *not* in existence today or the unknown market space. In blue oceans, "demand is created rather than fought over" [7].

The concept of the "blue ocean" quickly entered consulting and practitioner language and mind-sets.

Case Illustration: Cirque du Sol

One oft-quoted blue ocean success story is Cirque du Soleil, which disrupted traditional circus shows by borrowing ideas from Broadway. Notably, in an industry that was in steady decline, as alternative forms of entertainment lessened the appeal of a trip to the circus.

Cirque du Solei discarded many of the factors that had long ben mainstays of the traditional circus (most notably, the use of performing animals, with which newer generations were becoming increasingly uncomfortable).

The organization focused on reinventing, or disrupting, the industry by focusing on three factors: the tent (which they would glamorize), the clowns (but in a more enchanting, sophisticated style than the conventional slapstick model), and the classic acrobatic acts.

At the time of Cirque's debut in the early 1980s, other circuses focused on benchmarking one another and maximizing their share of the already shrinking demand by tweaking traditional circus acts. This included trying to secure more famous clowns and lion tamers, a strategy that raised circuses' cost structure without substantially altering the circus experience (and better lion tamers was hardly a compelling value proposition to younger generations). The result was rising costs without rising revenues and a downward spiral of overall circus demand. These efforts proved irrelevant when Cirque du Sol appeared.

Rather than following the conventional logic of outpacing the competition by tweaking the model, the organization essentially delivered the core function in a manner that customers now wanted. Moreover, in doing so they not only gained the custom of traditional circus goers, but also new customers—adult theatregoers.

Cirque du Soleil did not compete with established circus giants such as Ringling Bros. and Barnum & Bailey. Rather, it created uncontested new market space (Blue Ocean) that made the competition irrelevant. Significantly, the title of one of the first Cirque productions was "We Reinvent the Circus" [8].

The consequences? Cirque du Solei is now the largest theatrical producer in the world (10 million people watched their shows in 2016) while Ringling Bros. and Barnum & Bailey closed their tents for the last time in January 2017 and after 146 years. Their business model had failed to evolve—the once all-powerful sabre tooth tigers were extinct.

But there's a caveat here. Cirque du Solei has recently hit its own financial problems and is once again being reinvented, with plans to launch a theme park, a kids' entertainment project, and to design an interactive National Football League store in New York's Time square—and has aggressive plans for the Chinese market. Its mission: "….to invoke the imagination, provoke the senses and evoke the emotions of people around the world." Note, no mention of a circus.

Business Model Canvas

A further recent tool that is widely used is the Business Model Canvas. Initially proposed by Alexander Osterwalder, this is a strategic management visual chart and template preformatted with the nine blocks of a business model, which allows an organization to develop and sketch out a new or existing business model. It assists firms in aligning their activities by illustrating potential trade-offs. This is described through a model comprising key processes; key activities; key resources; value proposition; customer relationships; channels and customer segments; and the underpinning cost structure and revenue streams [9].

Brainstorming and agreeing upon the content of each of the nine components help an organization gain clarity around how it succeeds and initiates a rich conversation about the business drivers of the organization and, from that, becomes a useful contributor to discussions as to the strategic objectives that will ultimately house the corporate Strategy Map.

Capturing the Voice of the Customer

Increasingly, organizations are realizing the need to capture the voice of the customer as well as other key stakeholders throughout the strategy management journey. This is an important element of the strategic planning phase, so that all involved in the value chain are on the same page and understand the rationale of the strategy and the value delivered to each stakeholder group.

Tools such as customer co-creation, in which customers are actively involved in shaping solutions, play an important role here as well as in stakeholder analysis. We argue that this age of disruption, or the digital age, might also be named "the age of the customer."

Proliferating competition supported by ready access to masses of online data on potential product/service providers is gradually transferring all the power to the customer. There will be no going back on the supplier-dominated relationships of the yesteryears: yet, this is how most organizations still think, operate, and are structured. Consequently, we must more effectively involve the customer in the strategy development process and continue this involvement through execution.

That said, bringing customers to the heart of the organization has long been a corporate mantra, and terms such as "customer-centricity" and "delighting the customer" abound. However well intentioned, they are too often lost in the chaos of everyday service delivery and is often of secondary importance to maintaining margins over the shorter term. Moreover, too few organizations are taking the time to co-create product and service offerings with customers.

Co-creation requires working closely with customers (and potentially suppliers and employees) to redefine the value equation: genuinely talking to the customer, understanding the customer's experience, and finding ways to fix any problems.

Research Evidence

The 2014 Palladium research found that only 12% of surveyed organizations practice customer experience innovation, such as customer co-creation. This omission should be of serious concern to executives, given that just 3% report that they believe that satisfying the customer base is sustainable. Moreover, those organizations that are active in customer experience innovation are 12 times more likely to be market share leaders and 37 times more likely to be top quartile financial performers, according to the research.

Mapping the Customer Journey

To manage customer experience, a company must incorporate a structured approach, starting with the customer journey, a method to "step into the shoes" of the customers; understand their experience in the significant "moments of truth" with the organization; and understand their attitudes, behaviours, pain points; and so on.

Keep in mind that not all customers are the same, so perform customer journeys for each of the main customer segments. Customer journeys emphasize customer segmentation to ensure the best experience is designed for customers, according to their needs. For example, Vodafone identified "young, active, fun," as a customer segment. Such segmentation is based on smart data analysis and understanding the customers' preferences.

A graphical map (Fig. 2.2) shows the interactions that are meaningful for the customers from their point of view, incorporating their needs, expectations, and emotional experiences as they go through each step in the lifecycle. Combining customer emotions into a structured customer experience analysis is a relatively new approach and is one of the benefits of mapping the customer journey. By *understanding* customers from *their* point of view, organizations generate value—for the customer and for the organization.

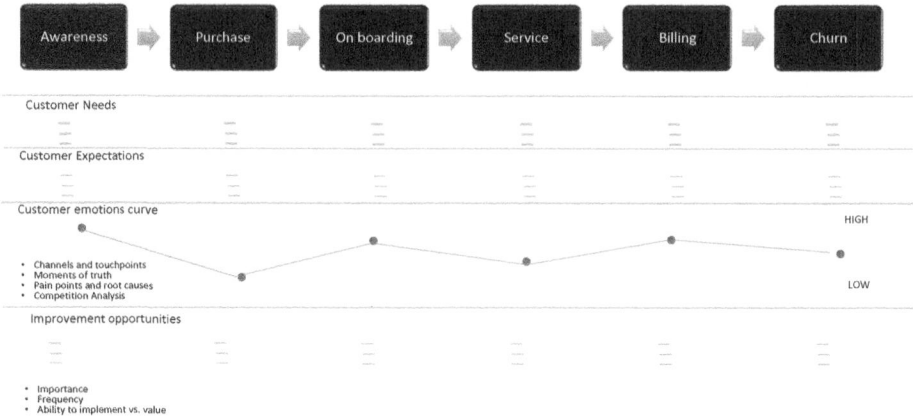

Fig. 2.2 Mapping the customer journey

Collaborative workshop methods that bring different parts of the organization together to create the map result in the alignment of all parties responsible for the customer experience—thus, breaking organizational silos and increasing innovative thinking.

Customer journey mapping serves as powerful input for useful improvement initiatives. These initiatives can be strategic or tactical and treated according to prioritization methods, such as by strategic fit to the objectives (see Chap. 5). Furthermore, looking at the relationship from the customer's point of view reveals innovative opportunities. For example, mapping children's experiences during hospital check-ups pointed to fear as a significant emotion and led to designing an MRI machine decorated with Winnie the Pooh. By reducing the fear factor (a huge value for the customers—children and their parents/carers), the hospital gained efficiency by reducing staff effort required to calm a child down, saving money on sedation drugs that were no longer needed, and shortening the idle time of the machine.

Emotional Touchpoints

With customer mapping, organizations are finding that not all customer experiences are of equal value, that there are certain "emotional touchpoints" that are the primary influences of customer behaviour and satisfaction, as identified by the children's hospital. In mapping customer experiences, it is important to identify these touchpoints. Yes, ensure that each step in the journey is satisfactory to the customer, but do the emotional touchpoints exceptionally well.

One European airline realized that what mattered the most to customers was being kept informed. Interestingly, they understood this when analysing which specific flights and routes scored the highest customer satisfaction scores. To their surprise, they found that the highest rating was from one route and on flights that had been delayed!

At first, this made no sense, but further analysis revealed that their staff kept customers well informed and were extra attentive before and during the actual flight on such delayed flights. The customers remembered these emotional touchpoints—not the quality of the food or the check-in process. See also Chap. 9: *Unleashing the Power of Analytics for Strategic Learning and Adapting*.

The "Unknowing" Customer

Although we must continually ask the critical questions, "what function or experience is the customer actually buying and how does the customer experience our world?" we must also bear in mind that customers might *not know* what they want or the experience they are seeking. Few people complained that the early generation cell phones couldn't take a photograph or connect to the Internet—but they readily bought it when they saw it. Steve Jobs realized that people did not actually want to buy a portable mechanism for making telephone calls, but a personal communication system.

This once again points to the danger of not keeping an eye on external happenings (be that through technology-based planning or more conventional approaches) and therefore our reliance on the measurement systems we deploy to monitor the success of execution. If we go back to just before Ford's incredible innovation and we surveyed the buyers of horses and carriages, we are sure the customer satisfaction scores would have been relatively high and the makers would have focused on the satisfaction scores to find ways to deliver a better product. What don't the customers like and let's improve it! They would then do so, their customers would be happier, their firm would make more profits and then—bang—they are out of business.

Case Illustration: How Alibaba Is Disrupting Retail

As a best practice example of disruption (which their competitors *should* have seen) consider the China-based Alibaba.

Founded in 1999 by a Chinese teacher, Jack Ma, whose ambitious dream was to implement e-commerce in the then very traditional Chinese retail market (certainly disruptive innovation). His central tenet was unusual: "Customers first, employees second, and shareholders third." In a letter he submitted before the IPO on the New York Stock Exchange in 2014 (the largest US IPO on record), he discussed his ecosystem-based business model, writing, "We believe that only by creating an open, collaborative and prosperous ecosystem that enables its constituents to fully participate can we truly help our small business and consumer customers. As stewards of this ecosystem, we spend our focus, effort, time and energy on initiatives that will benefit the greater good of the ecosystem and its various participants. We can only be successful if our customers and business partners are successful."

Has it been successful? Alibaba is the world's largest retail platform, as of April 2016, surpassing Walmart, with operations in over 200 countries, as well as one of the largest Internet companies. Alibaba generates more gross merchandise volume (GMV) than Amazon.com and eBay combined. Its online sales & profits surpassed all US retailers (including Walmart, Amazon, and eBay) combined in 2015. It has been expanding into the media and entertainment industry, with revenues rising by 3-digit percentage points year on year. As of October 2017, Alibaba's market capitalization stood at around $470 billion, placing it within the top 10 companies by market cap. in the world.

Note that Alibaba's mission is, "To make it easy to do business anywhere." Its vision (which we would argue is actually part of the mission) is, "We aim to build the future infrastructure of commerce. We envision that our customers will meet, work and live at Alibaba, and that we will be a company that lasts at least 102 years." The purpose and function are clear; the product or delivery mechanisms are not.

Parting Words

Professor Michael Porter has said, "strategy must have continuity. It can't be constantly reinvented." As much as anything, this speaks to the dangers of applying agile thinking and ideas to the complete management process, as this chapter has stressed. He has also said that, "a company without a strategy is willing to try anything," which we have also made clear a step on the road to oblivion.

But with that understood, where strategy starts to become more agile is in the setting of shorter-term and medium-term goals and their delivery mechanisms. We begin to discuss this in the next chapter.

Panel 1: Moving from Finance-Based to Technology-Based Planning

Technology underpins our personal and organizational lives, as well as that of a nation. Not only are we dependent on technology, the breadth and pace of technological change is accelerating.

Unfortunately, our ability to manage technology in a coherent and meaningful way in terms of creating measurable value on a repeatable basis has not evolved commensurately.

The generation of economic wealth, as an objective, is inextricably linked to the effective and systemized application of technology. Given this, in the digital age, it is imperative that organizations are accomplished at utilizing technology.

This is known as Technology-Based Planning (TBP) and is an approach through which all planning and decision making is centred on the acquisition and utilization of technology to create a competitive advantage. This is in terms of satisfying customer needs better than all the competition. By doing so, this dictates the amount of other resources that will be required and how they are deployed—R&D, innovation, funds, manpower/skills, natural resources, and so on. Key, therefore, is that planning begins with the following question: "What technology do we need so to create and maintain competitive advantage?" This is very different from "What resources do we have to invest in technology?"

With continuous, rapid, and extraordinary evolutions in technological and collaborative capabilities in this, the digital age, moving away from established twentieth century finance-based planning approaches to a technology-based approach is fast becoming a competitive imperative.

Two mechanisms underpin this:

1. The creation of a technology-space map®
2. An ability to navigate the technology-space

This ensures that organizations rapidly acquire and utilize technology with unprecedented speed, efficiency and agility to acquire, create and then maintain a competitive advantage.

A business adopting the automated innovation approach will not only have the potential to control a pivotal technology, but also a set of technologies in any present or future industry. Unlike traditional market-focused businesses, which generally rely upon a single pivotal technology for their strength, a business adopting an automated innovation approach will create an environment to enable a seamless transition from one pivotal technology to the next—within their industry or simultaneously across other industries.

The Myth of Disruptive Technology

Operating with the insights of the global technologyspace, all competitor technology manoeuvres are fully exposed well in advance, the notion of a "disruptive technology" can be seen developing many months before it is ready for the marketplace. From these insights, organizations utilizing technology-based planning quickly identify, track, and neutralize competitor "disruptive technology" initiatives before they can be executed in the marketplace.

Given this, organizations using the conventional finance and market-based approach are fully exposed to and exploited by organizations using technology planning.

The Myth of the Innovator's Dilemma
There is a myth called the "innovator's dilemma" concept, which says that successful companies may put too much emphasis on current customer needs, and by doing so, fail to adopt new technology or business models that meet unstated or future needs. The alleged innovator's dilemma for the business planner is then, "do we put our innovative resources on satisfying existing customer needs, or do we focus innovation on future breakthroughs that provide "disruptive innovation/technology?"

The reality is that to survive in the global marketplace, organizations must do both simultaneously—satisfy existing as well as future needs via technology-based planning and the delivery of a competitive outcome or superior performance. This is challenging to do via the conventional finance and marketing business model, but much easier via technology-based planning.

The innovators dilemma "only exists when planning is based on a conventional view of the marketplace which is completely unaware and totally blind to the technologyspace where all competitive activity is planned, developed and is clearly visible in advance of execution. The results of this will ultimately surface in the marketplace. Any business planner gaining insight into the technologyspace via technology-based planning capability would clearly see that there is no 'dilemma.'"

For organizations operating in the technologyspace, the ongoing competitive planning required for both existing (near-term) customer needs and mid and long-term competitive strategies are clearly exposed, enabling explicitly co-ordinated technology acquisition manoeuvres, with the objective of creating a sustained competitive advantage. There will never be a dilemma because the technology strategy created fully addresses "competitive advantage" in all time phases according to present, mid, and long-term existing and future customer needs.

Abridged from several LinkedIn blogs by Iain Wicking and based on the work of the US-headquartered Quadrigy in developing The Socrates platform [10].

From Finance-Based to Technology-Based Planning Self-Assessment Checklist

The following self-assessment assists the reader in identifying strengths and opportunities for improvement against the key performance dimension that we consider critical for succeeding with strategy management in the digital age.

For each question, any degree of agreement to the statement closer to one represents a significant opportunity for improvement (Table 2.1).

Table 2.1 Self-assessment checklist

Please tick the number that is the closest to the statement with which you agree		
	7 6 5 4 3 2 1	
The senior team of my organization generally believes the shaping of long-term plans is still an important requirement		The senior team of my organization generally believes the shaping of long-term plans is no longer an important requirement
The markets that we serve are significantly less predictable than what we saw five years ago		The markets that we serve are significantly more predictable than what we saw five years ago
The senior team has a consensual understanding of the term "strategy"		The senior team does not have consensual understanding of the term "strategy"
The sense of purpose of my organization is very well defined		The sense of purpose of my organization is poorly defined
In my organization, we regularly ask the "relevance" question: *Is what I am making or selling still relevant, and will it be relevant five or ten years from now?*		In my organization, we rarely ask the "relevance" question: *Is what I am making or selling still relevant, and will it be relevant five or ten years from now?*
The leadership team would be able to provide different answers to: What products and services do you sell? What do you sell?		The leadership team would not be able to provide different answers to: What products and services do you sell? What do you sell?
My organization regularly reviews the continued relevance of its business model		My organization rarely reviews the continued relevance of its business model
My organization is very good at innovating		My organization is very poor at innovating
Customer co-creation is used extensively when designing new/enhanced products or services		Customer co-creation is not used when designing new/enhanced products or services
We have a very good understanding of our customers' emotional touchpoints		We have a very poor understanding of our customers' emotional touchpoints

References

1. Jim Highsmith, et.al *Manifesto for Agile Software Development*, Agile Alliance, February 2001.
2. Michael Porter, *What is Strategy*, Harvard Business Review, November/December, 1996.
3. Michael Porter, *What is Strategy*, Harvard Business Review, November/December, 1996.
4. Henry Ford, *attrib*.
5. Vince Barraba, *The Decision Loom: A Design for Interactive Decision-Making in Organizations,*" Triarchy Press Limited, 2011.
6. Iain Wicking, *Linkedin Blog*, 2016.
7. W. Chan Kin and Renee Mauborgne, Blue *Ocean Strategy: How to Create Uncontested Market Space and Make the Competition Irrelevant*, Harvard Business Review Press, 2004.
8. W. Chan Kin and Renee Mauborgne, *Blue Ocean Strategy: How to Create Uncontested Market Space and Make the Competition Irrelevant*, Harvard Business Review Press, expanded edition, 2015.
9. Alexander Osterwalder, Yves Pigneur, Alan Smith. *Business Model Generation*, self-published, 2010.
10. Abridged from several LinkedIn blogs by Iain Wicking and based on the work of the US-headquartered Quadrigy in developing The Socrates platform.

3

Agile Strategy Setting

Introduction

In the previous chapter, we explained the importance of understanding the organization's sense of purpose. That is the reason the firm exists. Captured in a mission statement, this does not change often and indeed can remain the same for decades or even a century. This component of strategy management cannot be agile, and attempts to make it so will do little but suck out the essence of life from the enterprise. It will be left with no real meaning—no soul (Fig. 3.1).

Crafting a Vision Statement

Where strategy starts to become more agile is in the setting of shorter-term and medium-term goals and their delivery mechanisms. Here, being able to respond quickly to opportunities and threats becomes a competitive differentiator. The first step is to craft the vision of the organization. As with a mission statement, organizations often fail to capture the power of the vision as a strategic steer.

Not Visions: Advertising Slogans

A trawl through the recent vision statements of Fortune Global 100 companies uncovers some real gems: "To be the best consumer products and services

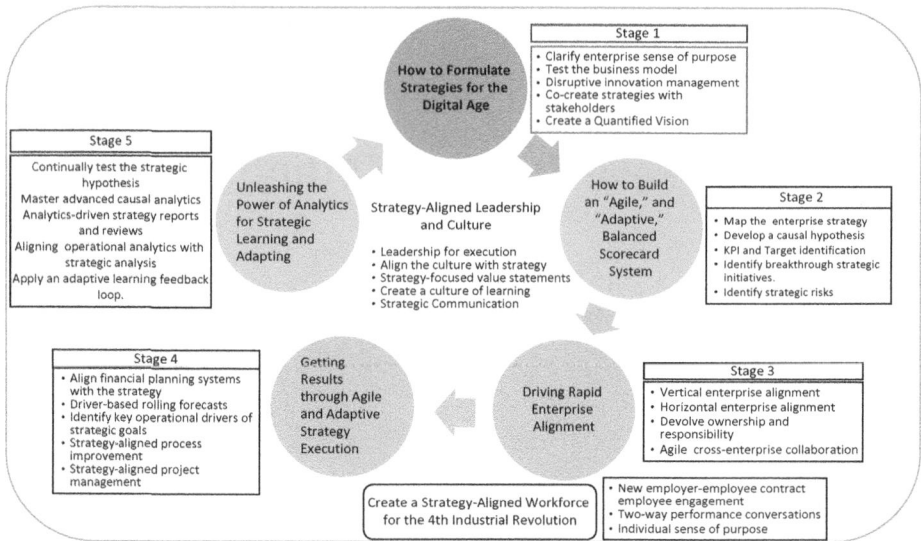

Fig. 3.1 Stage 1: How to formulate strategies for the digital age

company in the world," (Procter and Gamble); "To become a leading global energy company," (Lukoil); "Enel is the Leading producer and distributor of gas" (Enel); "To be the most competitive and productive service organization in the world," (Sociate General); and a real beauty, "No 1 LG" (LG).

Note that we are not picking on these companies specifically. Most of the vision statements that we reviewed are similarly vague and, to be blunt, of limited value. Such statements tell the reader nothing about what the organization wants to achieve over the timeline of the strategy, what success will look like, or how that will be achieved. What exactly does "a leading" company mean? Or to be "the best," and the mind simply boggles at trying to make sense of No 1 (in what exactly?). They are not visions—they are advertising slogans.

Moreover, some of the Fortune vision statements confused setting a goal and direction with values. For instance, "The continued success of Costco depends on how well each of Costco's employees adheres to the highest standards mandated by the company's code of ethics?" Applaudable, but hardly a vision.

Executable Visions

When crafting a vision statement, a critical question is "how will an employee (tasked with implementing the vision) understand what they are supposed to do differently to deliver success?" In our training workshops, we typically ask participants to state their organization's vision: most get on their smartphone to look it up. Hardly a sign that they find the words inspirational. Of course, it also means they are not being aligned to the vision.

The shaping of the vision statement is *the* most important step in the strategy planning process (and planning is different from the strategy setting steps described in the previous chapter). The vision is the anchor for all subsequent thinking—up to the identification of the strategic objectives, measures, targets, and initiatives that appear within a Balanced Scorecard system and throughout execution. The vision should precisely describe what the organization wishes to achieve through its strategy: the end goal.

A Quantified Vision

A well-crafted vision statement should be inspirational, aspirational, and, crucially, measurable (to capture whether, or not, the strategy has been successful). A "quantified" vision helps here.

Case Illustrations

Early scorecard adopter CIGNA Property and Casualty shaped a quantified vision to drive a major transformation exercise, which included a significant shift in market positioning and to turnaround horrendous financial losses. The vision simply and succinctly read, "To be a top-quartile [in profitability] specialist within five years." Thet were previously a generalist, so played in most insurance spaces.

The vision triggered, among other interventions, significant organizational restructuring. The firm, which had lost over a $1 billion over a three-year period, returned to profitability just two years post defining the new vision and implementing a scorecard [1].

As a further example, from the public sector (a University), "By 2015, our distinctive ability to integrate world class research, scholarship and education will have secured us a place among the top 50 universities in the world." Again, success indicator, niche, and timeframe are specified.

Mid-Term Visions

In short, a quantified statement moves the vision from a simple outcome statement ("to become the preferred supplier") to a more comprehensive picture of the enabling factors with which to achieve the vision. It provides clarity of focus and priorities.

A quantified vision is set for the lifetime of the strategy, so perhaps five years. Although appropriate, given the speed of change these days and the requirement for more agile and adaptive strategic manoeuvres it is often sensible to quantify a supporting vision over the mid-term: say, two years. This is consistent with the practice of mid-term planning, in which the mid-term plan provides more implementation details than the longer-term strategic plan and acts as connector to the annual plan. Typically, strategic initiatives are set over the mid-term, so this aligns well the Balanced Scorecard approach. It is over the shorter term that strategy comes alive, without losing sight of the guiding star that are the longer-term goals.

Moreover, five years is typically too extended a timeframe for employees to take particularly seriously: increasingly, a large percentage would have departed the organization by then and/or will believe that something will happen to completely change the organization before the years have elapsed (which is not uncommon these days).

Mid-term goals are more tangible and provide a greater sense of urgency. Furthermore, if the organization has a major transformational strategy, perhaps with very stretching financial goals that require significant organizational restructuring, geographical shifts in market focus, and new capability building, then the five-year vision runs the danger of simply being overwhelming.

> **Advice Snippet**
>
> A quantified vision statement includes three components:
>
> 1. A quantified success indicator (such as Cigna's top quartile in profitability)
> 2. A definition of one's niche (switching from generalist to specialist)
> 3. A designated timeframe (within five years).

Identifying the Value Gap

The quantified vision provides the base for identifying the "value gap," which is the difference between the organization's current performance and the quantified targets. Again, this can be done both over the lifetime of the strategy as well as over the mid-term.

The value gap becomes a powerful steer for planning and subsequent prioritization and resource allocation. For instance, say an organization wishes to "double EBITDA within five years," this might represent a $2 billion difference. The organization then needs to consider how to close the gap: via productivity or revenue growth interventions. It can then put a figure to each financial measure. As an illustration, for productivity improve cost structure by $750 million and improve asset utilization by $250 million, while for revenue growth expand revenue opportunities by and increase customer value by $500 million each. With the financial targets in place, attention switches to non-financial drivers and the interventions required to deliver these targets (Customer, process, learning and growth).

As a note of warning, from our experience, the early conversations around closing the value gap often centre on productivity goals rather than revenue, as the former requires considerably less creative thinking. Care must be taken to strike a sensible balance. Downsizing, and so on, are oftentimes too casually implemented.

Crucially, the value gap serves as the steer for setting KPI targets and defining the "performance gap,"—the difference between current KPI performance and the target, for both financial and non-financial metrics. Closing performance gaps ultimately closes the value gap and this, as we explain in Chap. 8, is primarily about developing "strategy-aligned" project management capabilities and implementing the identified interventions.

The quantified vision is also a key input into the Strategic Change Agenda, which we describe later in this chapter. However, there are important steps before this.

Environmental Scanning

Environmental scanning, in which an understanding of the external and internal worlds affecting the organization is, of course, important. Long established tools such as PESTEL (Fig. 3.2), Porter's Five Forces (Fig. 3.3), and SWOT (strengths, weaknesses, opportunities, and threats) (Fig. 3.4) are routinely used and have long been standard practices for most organizations.

An interesting spin on the traditional SWOT is a model developed by Mihai Ionescu of Strategsys. In this version, the identified strengths and weaknesses map more directly to the strategic choices (where to play, how to win) as determined by the strategic positioning, (Fig. 3.5). This evolution enables a more precise conversation on external world and therefore a smoother transition to developing the Strategic Change Agenda.

Many newer tools essentially do the same thing in understanding the linkage between marketplace dynamics and internal delivery models and capabilities. We describe some of the newer tools below, but this certainly not an exhaustive list—rather a flavour of what's available.

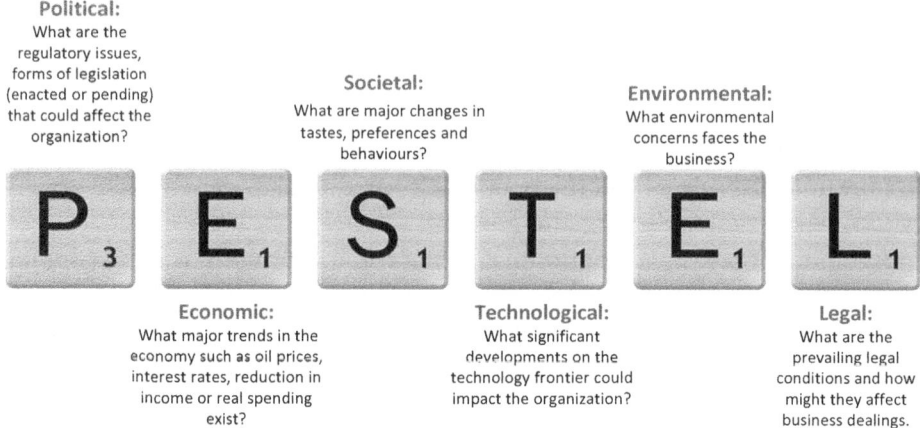

Fig. 3.2 A PESTEL analysis

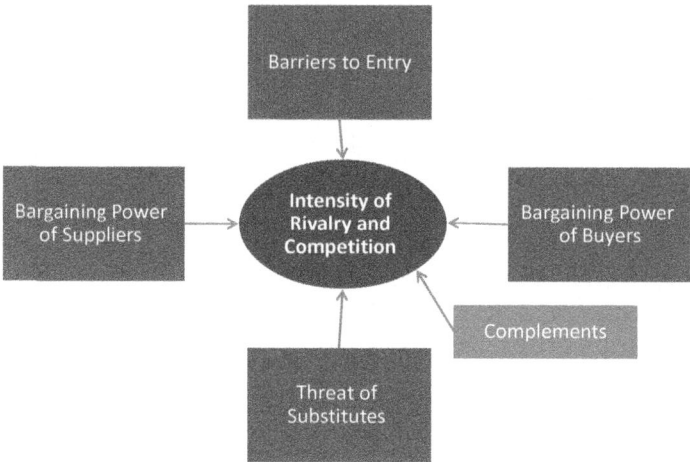

Fig. 3.3 Porter's Five Forces framework

However, whichever tools are used, we must move away from viewing them as once-a-year exercises. Scanning the external environment must be a continuous process.

Pattern-Based Strategy

The research and consulting firm Gartner has identified four disciplines to successfully adopting what they term a pattern-based strategy: pattern

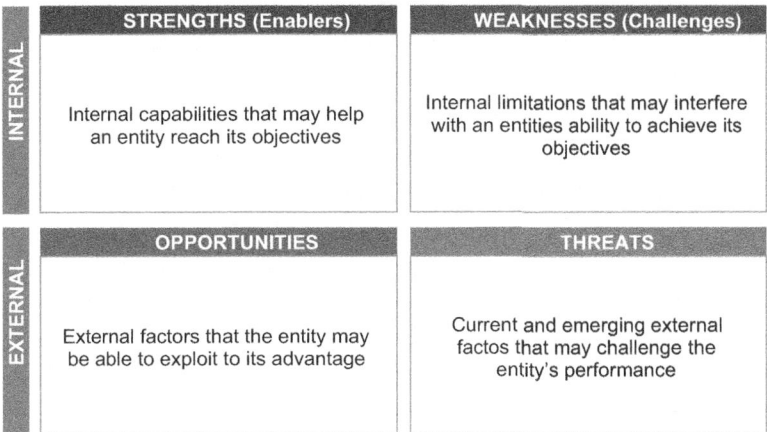

Fig. 3.4 A SWOT analysis template

seeking, optempo (operational tempo) advantages, performance-driven culture, and transparency.

Pattern Seeking

Pattern seeking comprises focusing on the competencies, activities, technologies, and resources that expose signals, which may lead to a pattern that will have a positive or negative impact on strategy or operations—focusing on those areas of vulnerability or risk and innovation/opportunity for the business.

Seeking patterns can mean looking inside or outside the organization and involves exploiting the new power of the collective—that is, exploiting collective knowledge with creative activities, and exploiting collective activities as an unexplored source of patterns.

"The collective is made up of individuals, groups, communities, crowds, markets and firms that shape the direction of society and business," said Tom Austin, Vice President and a Gartner Fellow. "The collective is not new, but technology has made it more powerful – and enabled change to happen more rapidly. The explosion of social software, for example, has enabled groups and individuals to rapidly form and rally to a cause, often resulting in significant societal changes" [2].

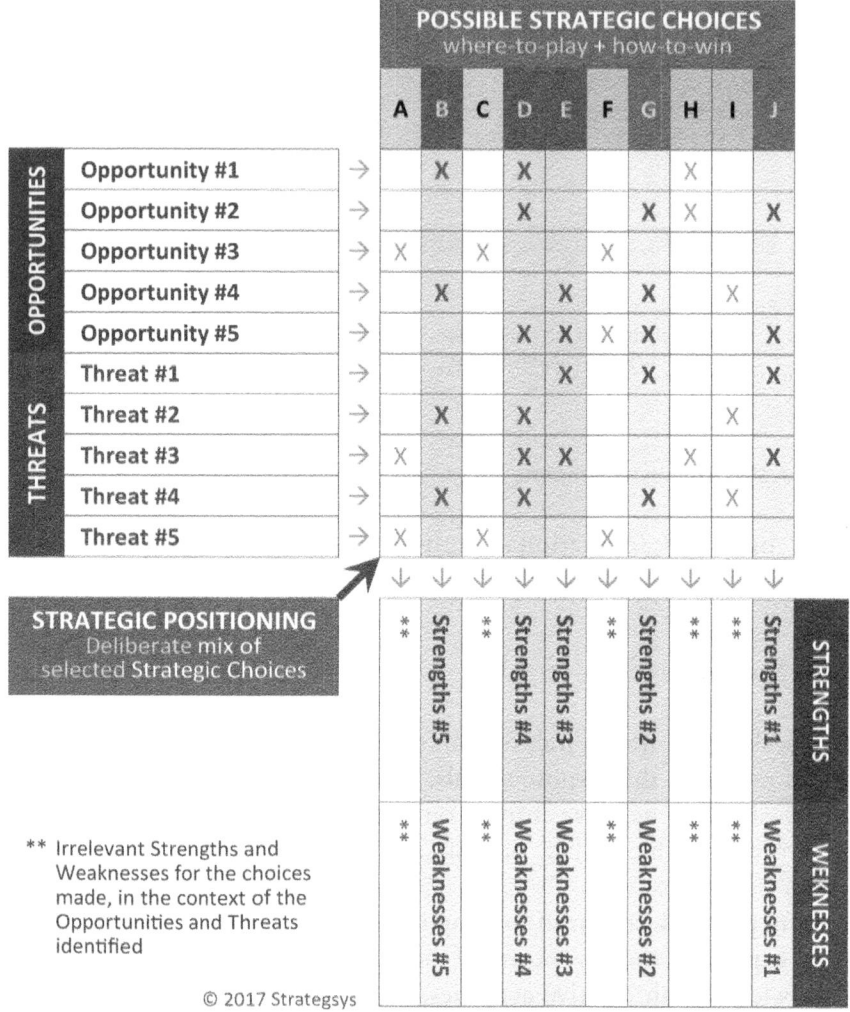

Fig. 3.5 Strategsys SWOT analysis. (Source: Strategsys)

Optempo Advantage

Gartner terms optempo advantage as representing the set of coherent guidelines and actions necessary for maximizing the allocation and utilization of enterprise resources as new patterns emerge.

In a pattern-based strategy, an organization excels by adjusting the relative speed of its operations better than its competitors do. However, enterprises must first understand the levers they can control to drive change. These levers are people, processes, and information. Business leaders can think of an optempo advantage as a formal management philosophy for improving their organization's competitive rhythm, so that it can match pace with purpose consistently and dynamically.

Performance-Driven Culture

A performance-driven culture enables an organization to monitor leading indicators of change. "Most organizations measure performance; they don't manage it. The traditional focus has been on measuring high-level, financially oriented outcomes after the event. This creates a reactive sense-and-respond mind-set," Sribar said. "Today's environment demands looking at leading performance and risk indicators to provide a forward-looking focus that then must permeate all levels of an organization—rather than just providing top-level measures. Changes in business strategy and operations will be reflected in changes in performance metrics, which will then drive change in behaviours and operational tempo" [3].

This component in captured within a Balanced Scorecard System, as we explain in subsequent chapters.

Transparency

In the context of a pattern-based strategy, transparency means both the demonstration of corporate health and the strategic use of transparency for differentiation.

If organizations can proactively evolve transparency from a once-a-quarter financial-results event to using it to set the right expectations of seeking new patterns and responding with consistent results, this proactive use might enable them to enter new markets, gain access to funds that competitors can't access, and demonstrate differentiation to customers and suppliers.

SCOPE Situational Analysis

Many commentators agree that SWOT (a tool originally developed in the 1960s) is somewhat out of date, although some state is still of some value. A more modern equivalent is SCOPE Situational Analysis.

To explain, the premise behind SCOPE is to offer a situational analysis that takes a more 360-degree view: encompassing past, current, and future perspectives as follows:

S – SITUATION: Rear-view—pertaining to conditions that have a relevant and material impact on planning decisions with regards to internal or external environmental factors.
C – CORE COMPETENCIES: Unique abilities or assets of the business that provide the basis for the provision and realization of value to customers and are critical to the creation of competitive advantage.
O – OBSTACLES: Potential issues or threats that could jeopardize the realization of the core competencies and thereby impinge on prospects.
P – PROSPECTS: Opportunities that exist internally or externally to the business that can enhance sales and/or profits, created through leveraging its core competencies and overcoming obstacles.
E – EXPECTATIONS: Future-view—predictions of future internal and external conditions that are likely to materially influence, positively or negatively, the delivery of plans to meet the identified prospects.

The Danger of Being Frozen in Time

The issue with the more static scanning models is that they are descriptive (a snapshot in time) and a once-a-year exercise, and once done, not thought about again until the next strategy review process. Markets, customers, and competitors do not freeze when the external environmental scan is complete, waiting to unfreeze just before the next scan.

Organizations must keep a constant eye on the external world and institutionalize this during execution, ensuring the capabilities are in place to respond to (and, even better, predict) marketplace shifts that might derail the strategic plan by altering the underlying assumptions and so weakening the strategic choices made for implementation. As Simbarashe Tembo, Head, the Strategy Advisory, East Africa, for the consultancy Ernst & Young comments, "the problem is once the analysis has been carried out, most strategists fail to connect the dots from this point onwards. Strategic choices, objectives, initiatives etc., become disconnected from the earlier work, such as the SWOT."

As we also explain in later chapters (most notably Chap. 9), retaining an external focus does not mean responding to every identified market movement. This will simply become a form of agile strategy firefighting.

There is still a need for a governance structure that allows for a formal review and reflection (but not just once a year). With a quantified vision set and value gap identified, a critical step here is capturing the insights of the leadership team through formal interviews (which provide critical information for the Strategic Change Agenda).

Senior Management Interviews

From our experience and observations, it is common for organizations to create a Strategy Map from scratch in a single senior management workshop and immediately post a SWOT analysis. Although seemingly valuable as it is time-efficient, this approach often leads to inappropriate outcomes. As an example, the map is shaped based on the views of the most dominant and outspoken members of the group. Conversely, the map captures the individual views of the whole team and so represents everything the organization does, not what matters strategically.

Although senior management workshops are critical (to debate and agree on the objectives, etc.), we recommend a preceding step of one-to-one interviews with the senior team, which are conducted post the environmental scanning. Based on the outputs of the previous strategy development work, these facilitated interviews are intended to tease out what each individual believes to be the key strategic outcomes and enablers, and what they personally think the organization does well and not so well.

Case Illustration: Norway-Based Power Company

A Norwegian-based power company realized the benefits of structured interviews (that were conducted by an external facilitator) when building a scorecard system in the mid-2000s. In creating the top-level Strategy Map and Balanced Scorecard, the in-house team adhered to the classic scorecard creation process.

The process began with the vision of where the organization wanted to be in the future and an analysis of the strategic plan. Based on structured interviews with senior managers, a draft Strategy Map was then created, comprising objectives for the four perspectives of financial, customer, internal process, and learning and growth. The senior team then debated and refined the draft map. A final map was then created with consensus.

In an interview with one of the authors, the then-CEO described a clear benefit of the structured mapping process. "For the new strategic plan, we created a Strategy Map. However, the senior team's discipline in discussing and agreeing the drivers of strategic success during the scorecard process meant we ended up with a Strategy Map that was very different from the one inserted in the strategic plan."

The agreed map then became the basis for a subsequent round of interviews and discussions to create the top-level Balanced Scorecard of KPIs, targets, and strategic initiatives.

Using an External Facilitator

It is possible that this facilitator is an in-house employee, but it is usually advisable (at least in creating the first map and scorecard) that an external facilitator is used. The reasons for recommending external support at the beginning of a scorecard program are myriad.

Firstly, a Balanced Scorecard system appears remarkably simple to construct. After all, it's basically a collection of strategic objectives with supporting KPIs, targets, and initiatives. (Does not appear that difficult to put together!). As a result, relatively junior staff with little knowledge of or experience in scorecard design and management are oftentimes tasked with its formation. This is a recipe for disaster and is why many scorecard efforts deliver little real value, and either wither and die or are reduced to KPI collection and reporting mechanisms that are effectively divorced from strategy.

An external facilitator should bring deep knowledge and extensive experience (but be careful here as this is not always the case—unfortunately, they too can be seduced by the "how hard can it be" belief).

The facilitator should know how to identify objectives and KPIs, how cause and effect works, as well as understanding how to overcome the significant challenges of rollout. In short, the facilitator should know what works, what doesn't, and the pitfalls to avoid: something that takes a long time to develop.

Furthermore, an external facilitator should bring independence and neutrality, melding together individual views without getting involved in politics or pushing their own agenda. This is a point that should not be underestimated.

It is rare for internal people to be able to remain completely outside of the politics of the organization. Either they do not have enough perceived authority to challenge the views of the senior team or, if senior, have their own strong interests or agendas. For example, a finance director might have the authority to lead a scorecard effort, but might push their own agenda and make the scorecard too finance-oriented (this is also true of IT or HR directors or even operational directors, as other examples). Moreover, a mid-level finance manager might be very reluctant to challenge the views of his/her boss in public (indeed, might see it as career suicide). Conversely, performance analysts might have the independence but not the authority required to influence the scorecard design process or to challenge senior managers.

Interview Questions

Senior management interviews should contain two or three questions for each perspective (or themes) and take no longer than one hour. Regardless of functional position, leaders answer the same set of questions. It is not the interviewer's role to enter a debate, but simply to collect, and then synthesize, responses.

Questions should be open and not closed and focus should be on the strategic priorities and capabilities. Crucially, it is important that, in these sessions, there is no discussion on KPIs or even initiatives. The danger here is that what is getting measured now, or what major initiatives are underway, will shape the choice of objectives, which is clearly back to front. This interview is about clarifying strategic priorities and identifying the capabilities that the organization has or must develop, and nothing else.

As a questions template:

- What are the financial/funder objectives?
- Who are the customers, and what are *their* objectives?
- What internal processes support the financial and customer objectives and what capabilities do we need to develop these processes?
- What culture, competencies, and technology support the internal objectives and what capabilities do we need to develop these processes?

The collated responses form the key input to the Strategic Change Agenda.

The Benefits of Anonymity

For interviews to be valuable, they typically should be anonymous. This ensures that key personnel can express their views in a safe and confidential environment. They can say what they really think! As one practitioner, interviewed for the book *Doing More with less*, stated, the individual and anonymous interviews proved, "a great opportunity for them to 'get things off their chest' in a safe environment and setting" [4].

Interview Outcomes

What tends to emerge from these interviews is a general, high-level understanding of the critical capabilities and relationships that an organization must master to deliver to the strategy. Where there is usually some divergence is regarding the priority capabilities and relationships—unsurprising, as functional heads typically see their own outcomes as well as processes and learning requirements as taking precedence.

Through one-to-one interviews, the external facilitator gains a clear picture as to how individual senior managers view the drivers of organizational strategic success—what they agree upon and where they differ.

Another benefit of having already done individual interviews, and cited by another practitioner interviewed for the book *Doing More with less*, was that, "as the external facilitator presented a consolidated view of the senior managers' views and not their individual responses, the leaders felt more comfortable in giving full and frank feedback, which strengthened the work" [5].

Based on these interviews, the facilitator prepares a first draft Strategic Change Agenda (see below). But note that, during this phase (and throughout the complete scorecard creation process), a small staff team (perhaps two or three in total drawn from different functions, who should be selected for their commitment to collaborative workings as much as their domain expertise) should be involved. This is particularly important when engaging an external consultancy, as the organization must be able to manage scorecards without external support once the consultancy period is over—this should be a contractual deliverable in any consultancy project.

Devolved Interviews

Purchasing Agency Case Illustration

As well as senior management interviews, it is sensible to secure input and comments from lower-level managers/teams. This is critical for securing buy-in from those that must actually implement the strategy (and thus overcoming the danger of building and mapping a plan and throwing it over the wall for others to execute). A good example of doing this well comes from the

North West Collaborative Commercial Agency (NWCCA), which is the collaborative purchasing agency for the UK National Health Service (NHS) trusts in the North West of England (Greater Manchester, Merseyside, Cheshire, Cumbria, and Lancashire).

When building the Strategy Map in 2009, three levels of interviews took place. At the senior manager/director level, focused one-to-one interviews were held to consider the entire strategy and to gauge each manager's take on why the organization exists, the priority areas in which it must excel, and the underpinning enablers of performance.

Facilitated focus group sessions for functional groups such as finance, operations, IT, and marketing followed. These sessions focused on what the functional areas must do to support the strategy as well as securing their views on mission-critical strategic goals, performance enablers, and so on.

A further interview phase involved selected external stakeholders as well as non-executive board members. Discussions centred on all objectives and outcome targets and not on the internal drivers of performance.

The data from the three interview streams informed the draft Strategy Map, which was finalized at a feedback workshop, attended by most of the interviewees. Those staff members that couldn't attend the workshops were invited to provide written feedback via e-mail. That said, the composition of the final map was the decision of the executive leadership team. This is right and proper, as they have ultimate responsibility for delivering the organizational strategy [6].

A Strategic Change Agenda

Something we have noticed over the years is that despite the growing number of sophisticated tools now available for strategy management, some of the most useful are the simplest; one of which is the Strategic Change Agenda, for which the senior management interviews are the key input. In our work with organizations, we encourage the spending of more time on shaping the change agenda than on choosing the objectives. From a well thought out Strategic Change Agenda, the objectives just fall out. They become blatantly obvious.

To explain, a Strategic Change Agenda is a framework to identify and assess current state—"As is…" and to project desired future states—"To be…" for key change/performance dimensions to which an organization must pay close attention in executing its strategy. These dimensions are organization specific and must reflect the firm's strategically critical capabilities, structure, and processes. Figure 3.6 provides an example.

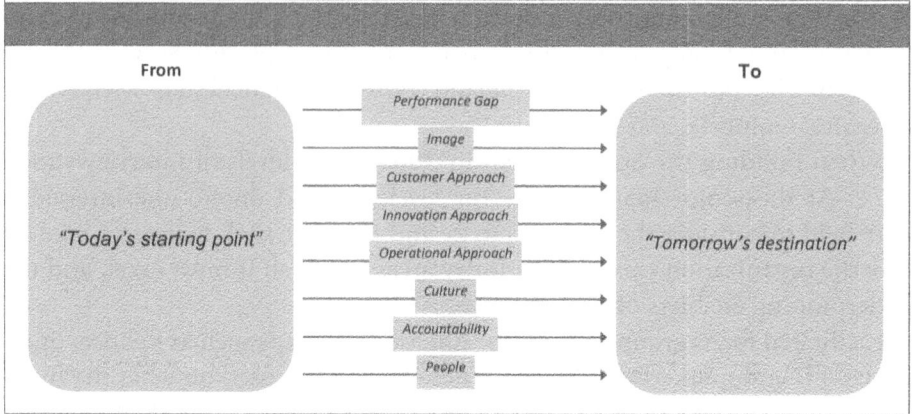

Fig. 3.6 Strategic Change Agenda

Case Illustration: FBI

As a best practice illustration, the US-based Federal Bureau of Investigation (FBI) developed a Strategic Change Agenda, which it called Strategic Shifts (with facilitative support from the then Balanced Scorecard Collaborative—which later morphed into Palladium) to reset priorities post 9/11 (Fig. 3.7).

Professor Robert Kaplan explained that to adapt to the new challenges, the FBI needed a brand new strategy and significant changes to the organizational culture. To do so, then FBI Director Robert Mueller recognized there was a need to educate the workforce as to the massive transformation required.

> The change agenda [which the FBI called Strategic Shifts] indicates that the agency would have to undergo a major shift from being a case-driven organization (reacting to crimes already committed) to becoming a threat-driven organization (attempting to prevent a terrorist incident from occurring).
> Dr. Robert Kaplan

He added that as a key strategic shift, agents had to make the mind-set and cultural switch from being secretive to working outside of traditional operational silos and to become contributors to integrated teams. "In even more of a discontinuity, the FBI had to learn to share information and work collaboratively with other federal agencies to prevent incidents that could harm US citizens."

The change agenda emerged from extensive dialogue throughout the organization, inviting all levels of the FBI to participate in setting the goals for the new strategic direction, which contributed to widespread understanding and

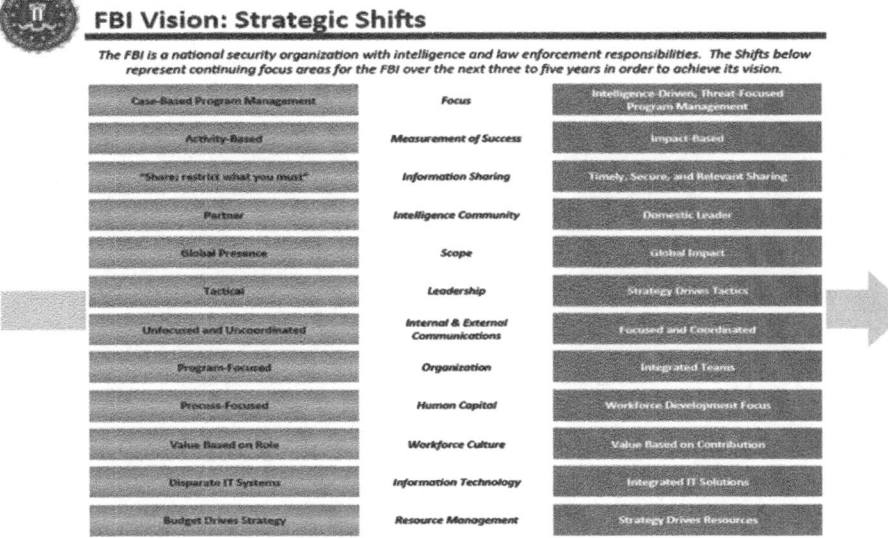

Fig. 3.7 FBI Strategic Change Agenda

support for the new strategy that followed. Kaplan explained that Director Mueller carried a laminated FBI strategic change agenda chart with him whenever he visited a field office. "If agents expressed skepticism about or resistance to the new initiative and structures he reminded them, using the single-page summary, why change was necessary." [7]

This is one benefit of a Strategic Change Agenda. It is a powerful communication tool, explaining how the strategy will be implemented and how far the organization is away from the desired state.

The Strategic Change Agenda helps the leadership team articulate and communicate the cultural, structural and operating changes necessary to transition from the past to the future.

Once the strategic shifts were formalized, the FBI's senior team built a Strategy Map and supporting Balanced Scorecard of KPIs, targets, and initiatives. As shown in Fig. 3.8, key strategic themes include *Management Excellence* and *Operational Excellence*—"Deter, Detect, and Disrupt National Security and Criminal Threats as well as Maximize Partnerships." Note how the objectives (and indeed the themes) link to the strategic shifts. As examples, the shift *the intelligence community* signalled a transition from the as is, (contributor) to the to be, (full partner).

This translated into the objective "enhance relationships with law enforcement and intelligence partners." The *scope* shift from domestic to global was

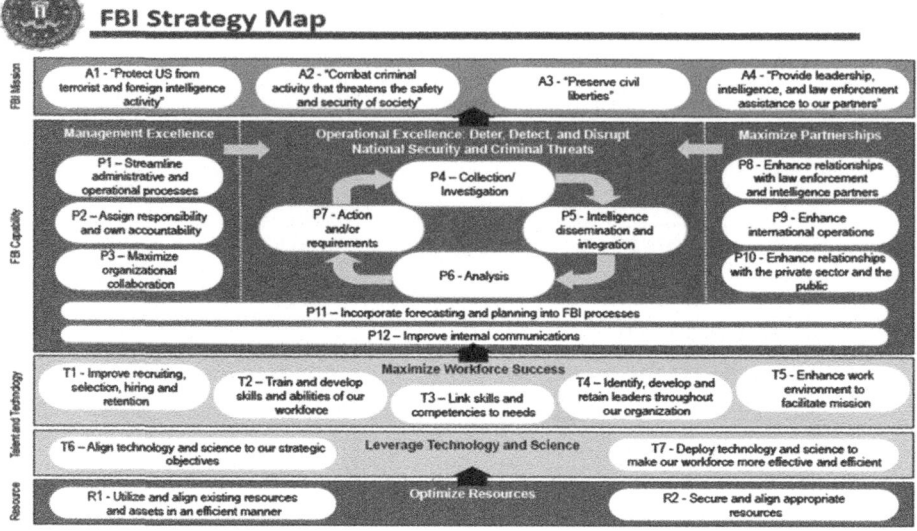

Fig. 3.8 FBI strategy map

captured through "enhance international operations." As a further example, the as is, for *information sharing* was "restrict: and share what you must," while the to be read, "share: and restrict what you must." This was actualized through the objective, "intelligence dissemination and integration."

Parting Words

With the understanding that the next step is to articulate the strategic objectives, we encourage organizations to arrange the change agenda according to the four perspectives of the Strategy Map (Fig. 3.9). This is a sensible approach. Once the performance dimensions, and the as is, and to be, states for the financial perspective have been agreed upon, the next question is, "what are the dimensions and the as is state required from the customers perspective to deliver those goals." Then, the same questions arise for the relationship between internal process objectives and customer, and learning and growth and internal process.

In addition to the value in identifying objectives, the rich conversations that happen here enable an initial understanding of the causal relationship between the perspectives. Also provided is clarity around where work needs to be done (in strategic initiatives and process improvements) to move from the "as is," to the "to be," and so close the value gap and deliver to the quantified

As Is	PERFRMANCE DIMENSIONS	To Be
	FINANCIAL	
	CUSTOMER	
	INTERNAL PROCESS	
	LEARNING AND GROWTH	

Fig. 3.9 Strategic Change Agenda, arranged from four perspectives

vision. Indeed, as many participants have said to us—and this applies to this stage, as well as to building the actual Balanced Scorecard System, "the quality of the conversations we have as a senior team are as important as the scorecard we develop."

In the next chapter, we discuss strategy mapping in detail.

Panel 1: War Games

A tool that is increasingly deployed is war games, which is useful to run when new strategies are being developed and when major strategic reactions need to be made (so a valuable counter to disruption). A war game helps uncovering hidden weaknesses—the organizations and its competitors—and so formulating best-course options.

A war game, which usually runs over 1–2 days, comprises three stages:

1. Determine who the key stakeholders are
   - Identify the key stakeholders
   - Develop the storyline: What are we going to simulate? What is the plan we want to test? (Alternative scenarios prepared for the war game session).
   - Detailed research and modelling of the environment and the behaviours of each stakeholder
2. Develop a briefing book for participants
   Develop a briefing book with the baseline of information about each stakeholder that will be represented in the game (pre-reading briefing book created and distributed in advance to the participants).

(a) Industry overview
(b) Stakeholders profiles (background, geographical footprint, business segments, market analysis, financial highlights and key indicators, strategy directions, etc.).

3. Prepare war game session
- Select the right people to play the game and brief them extensively on the role they will be play and the storyline.
- Identify core questions to be addressed through the game

A war game is run over several rounds and involves teams representing different key stakeholder groups: competitors, customers, regulators, suppliers, media, employees, the media, and so on. In round one, stakeholders are presented with the situation (typically one fact and date, such as "we are launching a new product line on such a date"). All teams prepare their response, then interact with each other to ask questions, lobby, partner, and the like.

Following a debriefing at the end of round one (and after teams have presented on how they will respond) the strategy can be adapted, and further rounds conducted until the most robust strategy is identified—and with alternatives available if things don't quite work out that way!

This is an important point, in today's world, in which we have passed through the technology "tipping point." Accurate predictions of what the future competitive landscape will look like are fraught with challenges. As one telecoms director said to one of the authors, "we struggle to figure what the world will look like six months from now."

Scenario planning is a similar, well-established tool for painting various "scenarios," of what the future might look like. To restate a key message of this book—in strategy execution, pay as much as attention to what's going on in the external world as to what's happening internally.

Panel 2: The OOLA Loop

In an interview for this book, John A Gelmini, Marketing, Innovation, Strategy and Partnerships Director, OS International Plc. (Consultancy, Innovation and Partnerships Division) described the value of the OOLA Loop as "for gaining future focus, optimize known risks and create temporary, 'windows of opportunity.'"

Gelmini explains that agility and the ability to metamorphose at speed need to support the unfolding strategy, so to cope with competitive shocks and "Black Swan" events (an event or occurrence that deviates beyond what is normally expected of a situation and is extremely difficult to predict).

"This involves incorporating the military doctrine of the OODA Loop, which emerged from military strategist and United States Air Force Colonel John Boyd's book '*A discourse on winning and losing*,' [8], which borrowed heavily from Sun

Tzu's *'Art of War,'* [9] and Iyamoto Musashi (c. 1584 – June 13, 1645) 'Book of 5 Rings.' [10] and was applied in Korea, Vietnam and the 1st Gulf War."

The phrase OODA loop refers to the decision cycle of observe, orient, decide, and act, which was originally applied to the combat operations process, often at the strategic level in military operations.

It is now also often applied to commercial operations and learning processes. The approach might be powerful for the digital age and the requirement, as we stress throughout this book, to meld strategy management into a single, integrated process and with a continuous learning look. With the OODA Loop, all decisions are based on observations of the evolving situation tempered with implicit filtering of the problem being addressed. The observations are the raw information on which decisions and actions are based and favours agility over power in dealing with human opponents in any endeavour.

The civilianized version of OODA Loop has four components:

- Observe—ideally without competitors knowing what you have seen and worked out.
- Orienting yourself and your position relative to what competitors are doing, again, without them realizing what your next move might be.
- Deploy means assembling your marketing collateral, salesforces, and social media campaigns, and readying them for action without their knowing what you are doing.
- Action—swift, precise, and unexpected. "The process is undertaken at speed which is varied to disorient the competitor so that he or she is reacting to events which have already transpired and is repeated again and again so the strategy is executed in a series of hammer blows which cannot be countered fast enough or at all" explains Gelmini.

Self-Assessment Checklist

The following self-assessment assists the reader in identifying strengths and opportunities for improvement against the key performance dimension that we consider critical for succeeding with strategy management in the digital age.

For each question, any degree of agreement to the statement closer to one represents a significant opportunity for improvement (Table 3.1).

Table 3.1 Self-assessment checklist

Please tick the number that is the closest to the statement with which you agree		
	7 6 5 4 3 2 1	
My organization has a well-quantified, long-term vision statement		My organization does not have a quantified, long-term vision statement
My organization has a well-quantified, mid-term vision statement		My organization does not have a quantified, mid-term vision statement
We have clear mid-term goals		We do not have clear mid-term goals
My organization is very good at external environmental scanning		My organization is very poor at external environmental scanning
My organization is very good at internal environmental scanning		My organization is very poor at internal environmental scanning
The annual environmental scan is a relatively quick process		The annual environmental scan is a slow process
We have clearly defined the "value gap" between present performance and future targets		We have not clearly defined the "value gap" between present performance and future targets
When updating the strategy, we gain input through anonymous interviews with senior executives		When updating the strategy, we do not gain input through anonymous interviews with senior executives
When updating the strategy, we create a Strategic Change Agenda or similar		When updating the strategy, we do not create a Strategic Change Agenda or similar

References

1. James Creelman, *Building and Implementing a Balanced Scorecard*, Business Intelligence, 1998.
2. Gartner Analysts, *Increasing Competitive Advantage Through Seeking, Modeling and Adapting to Emerging Patterns of Change*, Garner, 2009.
3. Gartner Analysts, *Increasing Competitive Advantage Through Seeking, Modeling and Adapting to Emerging Patterns of Change*, Garner, 2009.
4. Bernard Marr, James Creelman, *Doing More with Less: measuring, analyzing performance in the government and not-for-profit sector*, Palgrave Macmillan, 2014.
5. Bernard Marr, James Creelman, *Doing More with Less: measuring, analyzing performance in the government and not-for-profit sector*, Palgrave Macmillan, 2014.
6. Bernard Marr, James Creelman, *Implementing a Performance Scorecard for a Collaborative Commercial Agency*, Advanced Performance Institute, 2009.

7. Robert S. Kaplan, *Leading Change with the Strategy Execution System Harvard Business Palladium white paper,* 2015.
8. John Boyd, '*A discourse on winning and losing,*' 1987.
9. Sun Tzu, *The Art of War, Special Edition,* translated and annotated by Lionel Giles, El Paso Norte Press, 2005.
10. Iyamoto Musashi, *The Book of Five Rings.* Mass Market Paperback, 2005.

4

Strategy Mapping in Disruptive Times

Introduction

In the next two chapters, we describe how to build agile and adaptive Balanced Scorecard systems. Within this chapter, we explore Strategy Maps (the most important component of the system), while in the next, we consider the supporting scorecard of Key Performance Indicators (KPIs), targets, and initiatives.

As we stressed in earlier chapters, we need to be careful in applying agile thinking to all aspects of the strategy management process. However, it is more readily applied during execution (where work gets done) but oftentimes, adaptiveness is more appropriate.

Starting with the Strategy Map

Building a Balanced Scorecard System should always start with the Strategy Map, as this is the most important component. Moreover, if the earlier steps (quantified vision, senior management interviews, and Strategic Change Agenda) have been done well—see previous chapter—then the Strategy Map should be a very quick process.

A Non-Agile Process

Something we have noted is that oftentimes it takes an organization a long time (perhaps several months) to build the corporate map and scorecard and then even longer to build the aligned scorecard systems (see Chap. 6: *Driving Rapid Enterprise Alignment*). Of course, this is not helped by the need of many consultancies to maximize billable hours. An interesting spin on Albert Einstein's quote that, "The definition of genius is to make the complex simple."

In the digital age, it is simply ludicrous to spend an inordinate amount of time building and rolling out a Balanced Scorecard System. By the time the project is only half way complete, elements of the strategy might well be out-of-date or more likely some of the underpinning assumptions would be found to be wrong.

The scorecard becomes a barrier against agility and adaptiveness, which is not a good place to be. A corollary would be the annual budgeting process, which most managers believe to be out-of-date (and sometimes totally irrelevant) on the very day it is published (for more on this, see Chap. 7: *Aligning the Financial and Operational Drivers of Strategic Success*).

Actually, completing the senior management interviews, the Strategic Change Agenda, and the corporate-level Balanced Scorecard system is a very quick process. We're talking weeks, not months.

There are very good reasons why this should be the case. Firstly, prolonging the effort to build the system is often time and energy consuming. Once complete, there is little appetite to revisit (and especially if there's a sequence straight to a suite of aligned scorecard) until the following year. Hardly agile.

Secondly, as Hubert Saint-Onge commended in Chap. 1, a strategy is a set of assumptions that must be verified in execution—a set of assumptions! Much better to get to a scorecard that "feels" about right and then continually test and adapt it rather than try to configure the "perfect" system (which would almost certainly be incorrect, anyway).

Testing on a continual basis is something we explore in detail within Chap. 9: *Unleashing the Power of Analytics for Strategic Learning and Adapting* and discussed below.

Writing Objectives

Through our field experiences and research efforts, it has been obvious that many organizations make fundamental errors when wording the objectives. Let's go back to scorecard basics and the original questions posed by Kaplan and Norton.

What Does Success Look Like in the Eyes of Shareholders (or Funders)?

There are sometimes mistakes with the financial (shareholder) perspective. Although seemingly a simple collection of financial objectives, it might be sensible to ask shareholders what they want from the organization. It is, after all, what success looks like *in the eyes* of shareholders. In some cases, they will want more than short-term gains and be concerned with governance or sustainability, for example. Similarly, if creating a top-level stakeholder perspective in a government or not-for-profit (the funders) then ask them what they want to see as outcomes for the organization.

What Does Success Look Like in the Eyes of Customers?

When designing strategic objectives for the customer perspective, a common mistake is to create objectives that describe what the organization wants from the customer relationship: objectives such as increased customer loyalty or increased share of wallet abound.

The customer perspective represents outcomes that are of value *in the eyes* of the customer. Customers rarely procure a product or service because they want to be loyal, and how many ask how they can give their supplier more money?

Therefore, write the objective as if the customer said it, such as, "Reliably provide me with quality water at a reasonable price" or "Provide me with the highest quality of care in a safe and respectful environment" (Fig. 4.1 provides an example of a Strategy Map with corrected worded customer objectives).

A trap to avoid is automatically assuming, when writing the objective, that the customer is referring to your organization. When they say, for instance, "Be my preferred partner," it's important to realize that they are saying that they would like a preferred partner, but that does not necessarily mean it has to be your organization. In such an example, it's important to think through what they mean by preferred partner and to ensure that through the work done in the supporting enabling objectives they choose your organization over other candidates.

"To Satisfy Our Customers, and Stakeholders, in Which Internal Business Processes Must We Excel?"

This is the perspective where the wording is generally fine (unsurprisingly, as it is takes us into the world of operations, which most leaders are more comfortable with than strategy).

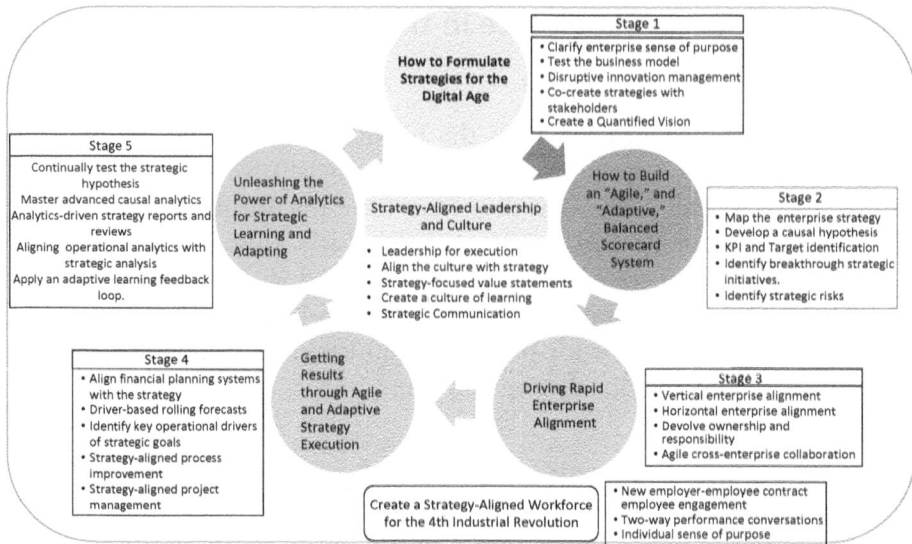

Fig. 4.1 Stage 2: How to build an "agile," and "adaptive," Balanced Scorecard System

To Achieve Our Goals, How Must Our Organization Learn and Develop?

Typically, the learning and growth perspective is the least well thought-out. The people objectives are usually generic and vague, such as "create a high-performance culture", "satisfy employees," "live the values," and so on. This holds true for the information objectives; "Support the organization with good information systems" being one common example. Again, organizations miss out on a golden opportunity to think through the people and technology capabilities required to deliver the internal processes to the expected level of excellence.

An organization that one of the authors worked with in 2017 had a set of learning and growth objectives that were very generic and, as they explained, not really of much value. Doing the change agenda led to a switch from an objective of "Recruit, develop and retain talent" to "Recruit talent to be [regionally]-focused," which is from where the significant market growth was to come.

Furthermore, keep in mind that the learning and growth perspective is *not* an HR perspective or an IT perspective; it is an enterprise-wide perspective. It is too easy to hand this section over to these functions "to sort."

Keeping Strategy Maps Focused

This brings us to another general mistake in Strategy Map development. There is a common belief that the map should reflect all areas of the business critical for making strategy: "everyone's everyday job." This often leads to maps that, and with little more than minor tweaks, are applied to any organization (a common gripe from managers that have to apply scorecards). Moreover, it oftentimes leads to "big" Strategy Maps of, say, 25+ objectives. Ideally, Strategy Maps should be limited to around 15 objectives and perhaps less.

Although appropriate for a Strategy Map that supports incremental growth in a relatively stable marketplace, capturing the whole of the organization is not useful if the firm is driving transformation change, with a dramatic shift in market or product focus.

This was the case with the example above of the organization looking to become regionally focused within one rapidly growing sector—and to support an extraordinarily rise in revenues over the timeline of the strategy.

Previously, this organization focused on a different region and on serving several sectors (albeit with similar technologies). The map they had originally developed tried to capture all parts of the business and, so, was very generic, which came out very strongly during the executive interviews held as part of creating the new map. Getting the leadership team to understand that this was not required led to the shaping of a map that focused on growth within the new region and in one sector. The CEO said, "We now have a map that we can truly call our own."

This is not to say the other work going on in the organization was not important, just not the current strategic focus. In such instances it might be useful to relabel the map as, in this case, a Growth in [region] Map. Good internal communications is required to ensure people understand that their work is still very important and indeed might be the next focus area. In addition, these parts of the enterprise must still align to the strategy and how to do this is something we consider in Chap. 6.

Objective Statements

Objective statements provide a fuller description of the objectives on the map (which due to the constraints of space are limited to a few words). At the outcome levels, the description details (typically in a single paragraph) what the objective will look like if achieved. At the enabler levels, it explains why the objective is strategically important (how it delivers outcomes) and achieved (what needs to be done) and two paragraphs will suffice.

Case Illustration: Hospital Complex

As an example, a hospital complex had an internal process objective to, "Assure service and process excellence & optimize the customer experience."

The objective statement reads:

> Key to our success is ensuring that from entering to leaving the hospital the patient's experience is as comfortable and stress-free as possible.
>
> *[This explains why it is strategic important in delivering the customer objective of, "Provide me with the highest quality of care in a safe and respectful environment that is easy to navigate."]*

> We will achieve this through implementing an end-to-end process that seamlessly integrates the process steps from the point at which the patient enters the hospital system, through the coordination of care whilst the patient is within the hospital, and finally to how the patient is discharged from the facility.
>
> *[This explains how the objective will be delivered and thusly the customer objective – cause and effect].*

In effect, the objective statements collectively tell the story of the strategy and the underpinning causal assumptions (and can be easily woven together to create a single value narrative). From these statements, value drivers are drawn, which makes the identification of KPIs a relatively straightforward task (see next chapter). They are also helpful in identifying strategic risks (see Panel 1 and the next chapter).

Consistent Interpretations

Another key benefit of objective statements is that they ensure a consistent interpretation of the objective as it threads its way across the enterprise. The fact is the senior leadership team, which agrees on the objectives, will have a shared understanding of the importance and meaning of the few words that describes them (or at least should).

But, outside of the leadership team, people will put their own interpretations onto the meaning—often function-centric, leading to misalignment rather than alignment. For instance, the senior team might have a clear understanding of the importance and meaning of "Develop a High-Performing Culture" but this will mean many different things to many people.

> **Advice Snippet**
>
> When writing an objective statement, keep the following guidelines in mind.
>
> - At the outcome level, just describe why the objective is strategically important—this represents the desired result
> - The enabler level describes why the objective is important for delivering to the outcomes as well as how the objective will be delivered (perhaps three components)
> - Complete the financial objective statements first, and then customer, internal process, and learning and growth. This is critical for visualizing cause and effect
> - Keep the statement short. One paragraph explaining the strategic importance and (at the enabler level) a second describing the how (which might be listed as bullet points)
>
> Pull the statements together to create a value narrative. This is valuable for communication purposes and for telling the story of the strategy in more detail than can be provided through a Map.

The Value of Strategic Themes

When thinking about alignment and the cascade of the scorecard system, it is worth considering the use of strategic themes, through which strategic objectives are grouped into several categories within a Strategy Map—such as customer management, innovation, operational excellence, and so on (as it is easier to align devolved functions etc., more directly to a specific theme).

In an interview with Dr. David Norton, he described the power of Strategic Themes. "Strategy is about delivering solutions to common challenges that the organization is facing, and this is at odds with a vertical structure," he explained. "By identifying and laying out strategic themes on a map, organizations are able to overlay a horizontal form of management onto the necessary hierarchical structure."

Designing Themes

Strategic Themes were originally positioned exclusively within the internal process perspective. The reason being because that is where work gets done.

Therefore, objectives were arranged according to process-focused themes, as shown in Fig. 4.2.

Together, these themes delivered the customer outcomes and from that financial results. However, organizations have since shaped themes according to their own needs and oftentimes across perspectives. Arrangements vary.

Figure 4.3 shows a government entity example of how themes cut across customer and internal process perspectives—in this case, attract targeted visitors, improve tourists' experiences, and contribute to social and economic development.

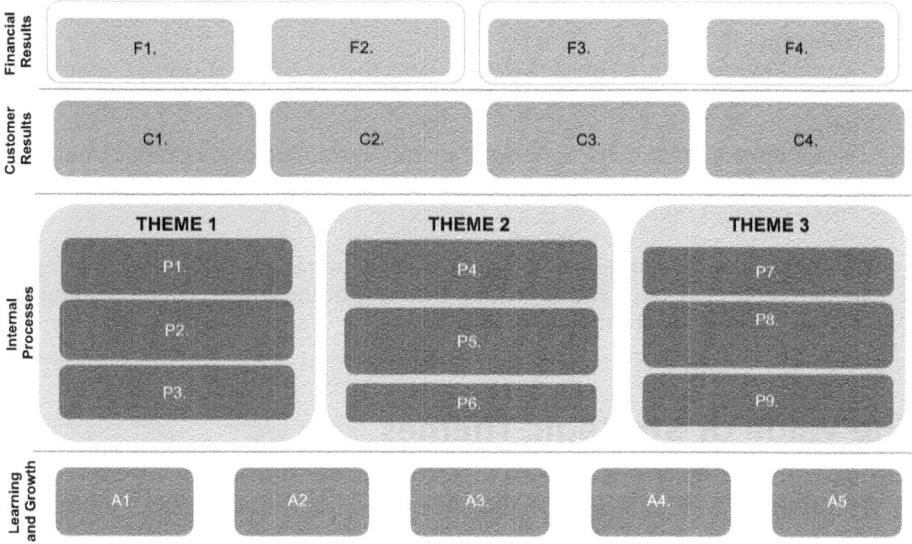

Fig. 4.2 Strategic themes within the internal process perspective

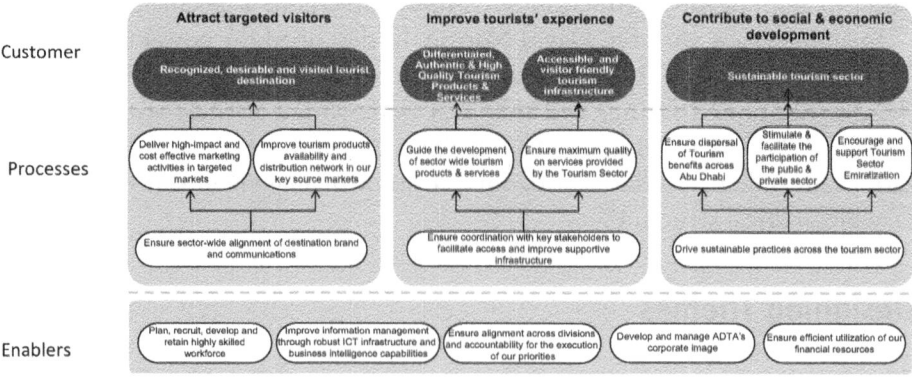

Fig. 4.3 Example of strategic themes within customer and internal process perspectives. (Source: Palladium)

Case Illustration: AW Rostamani

As another take, the Dubai-base AW Rostamani Automotive first identified six key pillars of a World Class Organization (that would deliver to their 2015 goal of being a $2 billion company) (Fig. 4.4). These pillars were transformed into Strategic Themes on the Strategy Map. The organization met their 2015 target by the end of 2014.

As an aside, its new corporate-level Strategy Map is designed in the shape of a car, something that one of the authors advised them to do as they were seeking a more impactful and inspiring way to use the map as a communication vehicle.

Constructing Theme Teams

Whatever the construct, a member of the executive leadership team must lead the "Theme Team." Only they have the power to drive the end-to-end process changes required to deliver transformational change, as well as the authority to second the resources required to work on strategic initiatives (functional heads typically balk at losing their "star" performers—the most likely candidates for large and potentially complex change projects).

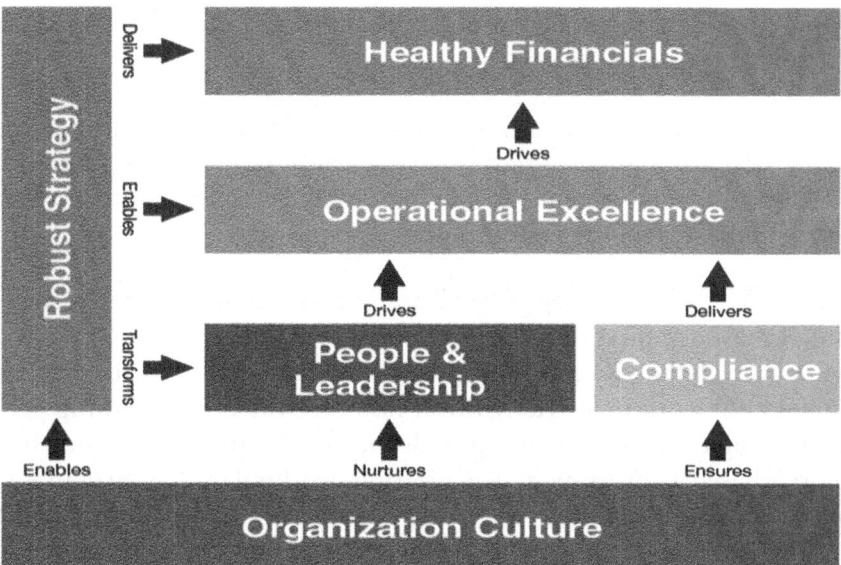

Fig. 4.4 A.W. Rostamani strategic pillars

A team of managers support the theme leaders and are responsible for the specific objectives within the themes—and supporting KPIs and initiatives.

Potential Downsides of Working with Themes

We must stress, however, that managing strategy through themes does come with some "watch outs" that might lead to them being performance-blockers rather than enhancers.

Although enabling executives to separately plan and manage each of the key components of the strategy, themes still need to operate coherently. That is, working together as an interdependent whole that *collectively* delivers strategic success.

A real danger is that themes default to into new, function-style silos, in that the members of the team focus only on their theme and ignore the others on the Strategy Map: encouraging battles for scarce cross-enterprise resources between theme owners.

Sandy Richardson, Founder and CEO of the Canada-based consultancy Collaborative Strategy, explains how a strategic theme may perpetuate a siloed approach to business thinking and operations. Strategic objectives are ideally cross-functional in nature, and cause and effect relationships exist across an organization's entire strategy map. The goal in strategy implementation and management is to leverage the Strategy Map and get the organization working together cross-departmentally on, and thinking more broadly about, the paths to mission and vision achievement.

> While asking a sub-team to manage a 'mini' Strategy Map, focused on a specific theme area, may ensure the successful achievement of the strategic theme, it may also result in sub-optimizing the achievement of other strategic theme areas and/or the entire business strategy.

Frederick W. Taylor's message (see Chap. 1) is well ingrained in the subconscious of most organizations.

Moreover, and as Richardson says, cause and effect, or causality, typically does not work in a strictly linear fashion, which is something people often lose sight of when working with themes. As an example, one of the authors advised a government organization, working in a country in which contractor treatment of low-wage migrant workers was a significant issue, that the "partner management," objective within the "Outsource and Deliver" theme directly impacted the "national and global reputation," objective within the "Customer Excellence" theme. As they were located in different themes the causal relationship was easy to overlook. Not managing this causal relationship carefully could have led to devastating consequences.

The Power of Cause and Effect

Something that organizations usually overlook, or to which they at best give perfunctory attention, is the causal relationship between the objectives (and indeed elements of the statement).

Most Strategy Maps that we see are not maps, but simply a collection of objectives—useful for communication purposes and for providing a succinct view of the strategy, but not for visualizing, and from that, testing the underpinning causal assumptions. A splattering of "arrows" throughout the map purportedly demonstrate a causal relationship. When reviewing Strategy Maps, we ask leaders to explain their hypotheses as to how improvements to a learning and growth objective will eventually deliver financial (or stakeholder) outcomes.

Normally, the answers are somewhat vague and hesitant. We will then ask for a written description of the causal relationship. If the map is collocated per strategic themes (as examples, customer focus, operational excellence, and innovation), we ask for a description at the theme level. This strategic performance narrative proves a powerful mechanism for leaders to gain a better understanding of their hypothesis and the causal assumptions (and is a critical purpose of the objective statement). Historically, cause and effect analysis has not been a standard practice of most scorecard users. The "arrows" placed on maps are rarely validated.

To be fair, there are good reasons why, historically, the required attention has not been given to causality. We simply did not have the organizational capabilities (people skills and software) to do this work. Or rather, it could be done by hiring a boatload of very expensive statisticians—rarely feasible. Advanced data analytics make this task relatively straightforward and financially viable (see Chap. 9).

Furthermore, the need to understand causality is precisely why weightings should not be used in a Balanced Scorecard system. This is a long-held argument in the strategy execution community, and we present our arguments against weightings in Panel 1.

Age 2 Balanced Scorecard Systems

Within Chap. 1, we outlined Dr. David Norton's thoughts regarding Age 2 Balanced Scorecard systems. Age 1, he argues, was essentially about building scorecards, communicating strategy, and alignment. All of these are still important and the foundation for Age 2.

Causal analytics is one of the components of Age 2 systems. Although we discuss analytics in detail within Chap. 9, where we explore some of the advanced data analytics tools that can now be used to gain better insights into the performance of the strategy (big data in particular), it is important to discuss here—for one critical reason. Perhaps for the first time since Strategy Maps were introduced to the strategist's toolkit, we are finally able to realize the promise of the map—to describe, manage, and test cause and effect.

The Contribution of Early Pioneers

Although advanced data analytics is enabling causal understanding, it is not an Age 2, or digital-age, evolution. The map's value for doing this was precisely why Strategy Maps evolved out of the first generation Balanced Scorecard—that did not include maps.

Strategy Maps emerged when early scorecard pioneers such as Cigna Property & Casualty (P&C) and Mobil Oil's North American Division found it valuable to separate the objective dimensions of the Balanced Scorecard framework to lay out a cause-and-effect relationship between the objectives housed within the four perspectives: typically, enablers (learning and growth and internal process) and outcomes (customer and financial). Originally called "Linkage Models," the idea of the Strategy Map was born. Figure 4.5 shows the Linkage Model for Cigna P&C.

The value identified by the pioneers was in laying out a *cause-and-effect relationship*, thus, understanding how progress to one or more enabler objectives delivered the desired outcomes. Indeed, it transformed the Balanced Scorecard System from a "balanced" measurement system to a full-fledged strategy execution model. Dr. David Norton has said that, "The idea of the Strategy Map was as important an insight as the original Balanced Scorecard framework."

However, analytical tools are only of value if there's an understanding of the causal assumptions. Organizations require the discipline of ensuring that when formulating objectives (and this starts with the Strategic Change Agenda), cause and effect is at the front of the mind. Explain how these

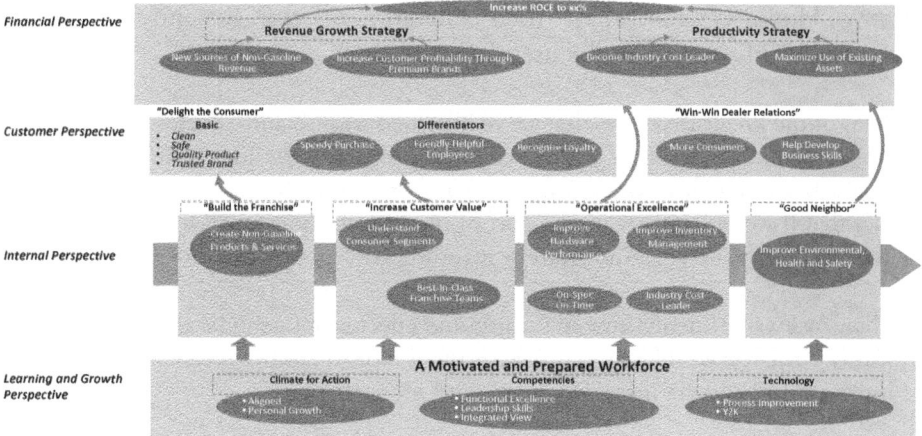

Fig. 4.5 Mobil oil linkage model for 1994

customer objectives will deliver the financial results, how the internal process objectives deliver for the customer, and—the one relationship that is typically not considered well—how the learning and growth objectives deliver the internal processes.

There's an important caveat here: do we accept the premise that improving capabilities within the learning and growth perspective leads to the effective delivery of strategic processes and from that the realization of customer and financial outcomes? Yes, we do! Do we think that the causal logic of a map accurately describes how value is created in an organization? Well, not exactly. This requires some explanation, which leads to the wandering into the world of mechanics. Specifically, the differences between classical and quantum mechanics and how this relates to our understanding of how Strategy Maps work.

Classical Versus Quantum Mechanics

Classical mechanics tells us (among other things) that if we pull a lever in one place a predictable result will occur elsewhere. The basics of the laws of cause and effect. Quantum mechanics (admittedly at the level of the particle) tells us that for any initial situation, there are many possible outcomes and effects, with different probabilities: cause and effect is not deterministic—we cannot predict an effect, just calculate probabilities.

Back in 2006, one of the authors wrote a book called *Reinventing Planning and Budgeting for the Adaptive Enterprise* [1]. For that work, he interviewed

Jeremy Hope (1948–2011), then Director of the Beyond Budgeting Roundtable. He provided this useful observation:

> A lot of the literature shows that we've inherited a cause-and-effect, central control form of management from Newtonian physics. The idea that we have a clockwork mechanism where there are levers within levers and that we can predict that if we pull a lever in one part of the organization it will have a predictable effect elsewhere… And strategy mapping reflects that to some degree.
>
> There's a lot of evidence now that cause and effect is not a model that represents reality. "The real world is an eco-system and is more quantum mechanics. Things move around and are inter-dependent and more chaotic."

He added that, "What we're seeing now is a movement away from the Newtonian model to one that is more organic, more devolved, more adaptive and more flexible to change."

Hope's comments further underline the dangers of applying agile software development thinking to strategy. Predictability is an issue.

The Intellectual Capital Model

At the time, Jeremy Hope's argument reminded the author of the work of Hubert Saint-Onge (who was interviewed for this) and others in creating the Intellectual Capital Model in the late 1990s (Fig. 4.6). The argument was that value (both financial capital and enhanced non-financial capital) is created when customer, structural, and human capital interacted (as captured in the intersections between the capital components). It is the interactions that cre-

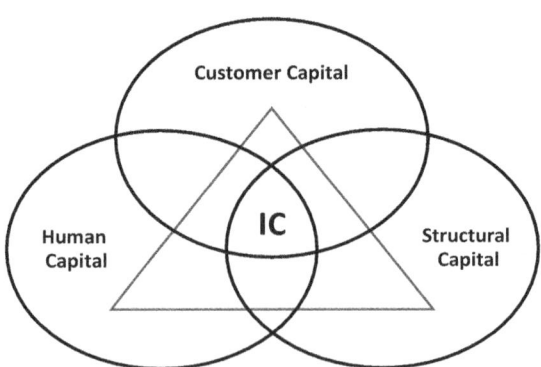

Fig. 4.6 Intellectual capital model

ate value. This makes it an imprecise exercise to measure the contribution of capital components in isolation, or as part of a logical casual sequence. This is certainly more of a quantum than a classical positioning.

That said, the actual drivers of the value created is not easy to isolate. Cause and effect is at best implied. However, although a useful model for visualizing value creation, such a model is extremely difficult to put into practice.

Interactions Between Intangible Assets

In the work of Doctors Kaplan and Norton, we also find evidence of value being created through interactions. Within the Kaplan and Norton Balanced Scorecard Certification Program (which both authors have delivered as trainers) it is taught that intangible assets do not deliver value in isolation. Specifically:

- The value of an intangible asset is influenced by its interaction with other intangible assets.
- It is difficult to isolate the value of one asset.
- The value of an intangible asset is determined by its impact on other variables.

These observations relate to the intangible assets found in the learning and growth perspective, but it is hardly a stretch to extend to process and customer assets (as described in the Intellectual Capital model).

Parting Words

In an interview with Dr. Norton, he stated that when it comes to describing how value is created, a Strategy Map sacrifices some degree of precision for practical usability (a 2D map rather than a 3D model) and that this was a conscious decision. "That organizational value is not quite delivered according to the flat cause and effect assumptions within a map, but the relationships provide a strong enough picture and narrative to overcome the negatives of imprecision." A recognition, perhaps, that value creation is as much about quantum as classical mechanics and that both have a place in managing strategy.

The fact is that to manage performance effectively, we need structure, process, and, not least, a roadmap to follow. A Strategy Map provides this performance management framework very well—and, in doing so, conforms to classical mechanics. Beleaguered managers require such aids.

Panel 1: Why Weightings Should Not Be Used in the Balanced Scorecard

For more than two decades, there has been a continued debate on the rights and wrongs of using weightings on a Strategy Map and Balanced Scorecard. Passions run high on both sides of the divide.

We take a firm and thus far unequivocal stance with regarding map/scorecard weightings. They should never be used. The reason—they make zero sense. And here's why:

Where Do We Place the Weightings?
A Strategy Map describes the causal assumptions regarding the capabilities that work together to deliver ultimate strategic success. Put simply, objectives within the learning and growth and internal process perspectives enable successful outcomes within the objectives at the customer and financial (or stakeholder) level. If that is accepted, then where exactly do we place the weightings?

At first glance, it might seem logical to give the highest weightings to the financial objectives; after all, these are the ultimate measure of a commercial organization's success. However, in keeping with the logic of the Balanced Scorecard system, these are simply the result of success within the other perspectives. So, it makes no sense to place the highest weightings here, as the results cannot be delivered without succeeding to customer and enabling objectives. Similarly, customer objectives are the outcomes of what happens within learning and growth and internal process: so, putting the highest weightings here is equally nonsensical.

Therefore, the conclusion might be that we should place the highest weightings on enabler perspectives, as this is how success is driven. But, what precisely is the purpose of having great scores for enabler objectives if the customers are leaving and the financials are nose-diving?

An Overall Scorecard Score!!!
Many organization like the idea of coming up with an overall "scorecard score." We struggle to see the point of this. Balanced Scorecard managers have said to us, and with great pride in their voices, things like, "I have a score of 90 on the scorecard." We have absolutely no idea what that means!

How to Prioritize Without Weighting
Of course, executing strategy is about prioritization. We do not argue against this, but weighting is not the way it is done. Prioritization should be through the resources that are committed to the designated strategic initiatives and process improvements, as this is how performance is enhanced.

Prioritizing Through Themes
This is one reason why themes are so useful on a Strategy Map. If, for example, there's a requirement to pay attention to the cost structure, then the bulk of the investments might go to initiatives and process improvements supporting an operational excellence theme. Conversely, if revenue generation is a priority, then the investments might be directed to an innovation theme.

In fact, at the outset of the financial crisis, some banks kept their Strategy Maps unchanged. They did not apply weighting, but did divert investments from growth to cost management themes and interventions. What these firms concluded was that, despite the severe pressures they were experiencing, the strategy was still valid and that, over the longer term, all themes/objectives remained of equal value. Just the short-term focus changed: after a while, the focus shifted.

Even when there isn't a "crisis," or an urgent requirement, due to funding issues, organizations still need to prioritize their investments over the lifetime of the strategic plan. This often means focusing more attention on certain themes/objectives than on others for a specified time frame. There is nothing wrong with this, and the sequencing of these investments is simply good management.

Weighting and KPIs
Furthermore, neither should we use weightings to prioritize KPIs. We argue that this suggests there has not been enough rigour in selecting the measures that most accurately assesses how well the organization is delivering on its objectives. So, the solution for too many organizations is to throw many potentially useful KPIs together, arbitrarily assign weighting, and come up with a collective score. This approach provides little insight into the drivers of performance improvement—but does create a lot of confusion.

Moreover, if the average score is green on a traffic light system, precisely what information does this provide into the drivers of performance? (See also the next chapter, where we discuss KPIs in detail).

A while back, one of the authors listened to Professor Robert Kaplan make a presentation on aligning incentive compensation to the Balanced Scorecard. At one juncture, he casually commented (more as an aside than anything else), "By the way, this is the only time weightings should be used in a Balanced Scorecard implementation." We can't argue with that.

Panel 2: Identifying Strategic Risks

There have always been risks attached to strategy execution, but with so many variables in play these days, it is ever present. In this the digital economy, successfully implementing strategy is about keeping one eye on performance and the other on risk. One without the other is not enough. Conventional Balanced Scorecard systems only consider the performance view, with Key Performance Indicators (KPIs) providing the evidence of success (or otherwise) of strategic objectives.

Risks Do Not Belong on a Strategy Map
Some organizations will claim that they cater for risk through its inclusion on a Strategy Map—as a theme, or even an objective. We've even seen it captured as a separate perspective. However, although common in the early days of the scorecard, we agree with Dr. Robert Kaplan, who said to one of the authors in an interview that, "risk management would become a strategic theme that would appear on the Strategy Map, alongside other themes such as customer service management and operational excellence. I now advocate that risk should not be on a Strategy Map at all, be that as a theme, perspective or objective."

He explained that the Balanced Scorecard is about managing and delivering performance, not mitigating risk, and that risks impact every objective on the Strategy Map, both financial and non-financial.

However, as a caveat, if the organization has a strategic need to develop risk management capabilities, then this might appear as an objective within the internal process perspective. This is very different from managing risks.

To manage risk effectively, we need a definition. "A key strategic risk is the possibility of an event or scenario (either internal or external) that inhibits or prevents an organization from achieving its strategic objectives" [2].

Note that a strategic risk event is a tangible occurrence. It is something that happens. Staff turnover is not a risk event, as turnover is an everyday reality of any business, although a defined loss of capability against a strategically critical skill might well be.

Although there are various ways to identify key strategic risks, one useful technique is to pose a \Key Risk Question (KRQ). For example, "What circumstances might lead to a degradation of processing accuracy?" might be a KRQ for the strategic objective "improve application processing accuracy," in a financial services company. A key risk event might be described as "the risk of a failure to achieve standards of processing accuracy caused by the loss of key staff resulting in the deployment of inexperienced staff."

Note that the strategic risk event is articulated as "the (key) risk of (what, where, when) . . . caused by (how) . . . resulting in . . . (impact)."

The Risk Bow-Tie

A popular tool for identifying what might potentially cause a risk to happen (such as lack of policies and procedures, inadequate activity management or external events) and the consequences should it materialize (direct, indirect, or intangible) is a "Risk Bow-Tie" (Fig. 4.7).

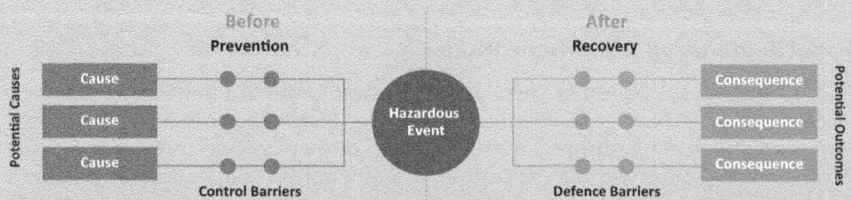

Fig. 4.7 The risk Bow-Tie

When using the Risk Bow-tie, start by focusing on events that could prevent the achievement of the strategic objectives. Once these are listed, start to develop as many potential causes as possible that will lead to the event happening and therefore the risk materializing.

Creating this long list of causes will help clarify thinking about the risk and form the base of a consolidated list of causes that are documented alongside the risk. This process is repeated for consequences.

Risk Heat Maps
With the strategic risk events identified, we sequence to assessing whether that risk will materialize and the effect on the organization if it does. This assessment often begins with a "Likelihood and Consequence" matrix. This simply plots on a vertical axis the likelihood of a risk materializing and the consequence to the organization if it does. The point where likelihood and consequence meet determine the risk's position on the matrix (the Risk Heat Map) and therefore the level of urgency for risk mitigation.

Four Perspective Risk Map
While the Risk Heat Map is a well-known tool, one innovation (pioneered by Andrew Smart, CEO of Ascendore and explained more fully in a 2013 book by Smart and Creelman) is a Four Perspective Risk Map [3]. This brings key risks together, enabling their visualization in relation to each other.

The Four Perspective Risk Map enables organizations to focus on risks in each perspective and explore the relationship between risks across perspectives and to identify risk clusters. For example, one organization that uses the Four Perspective Risk Map focuses attention on the risks within the outcome perspectives—financial and customer—as a starting point for their monthly risk review. The senior team explores the causal relationship between objectives, using both the Strategy Map and Four Perspective Risk Map, believing that taking this approach enables them to manage and monitor the delivery of their strategy.

The Four Perspective Risk Map helps senior teams to answer the questions:

- What level of risk are we taking?
- What level of risk are we exposed to?
- What are our main exposures?

Just as KPIs track performance to strategic objectives via a scorecard, KRIs monitor exposure to key risks on a "Risk Dashboard" (we prefer the term dashboard, simply to differentiate from the performance-focused scorecard). We discuss KRIs in the next chapter.

Self-Assessment Checklist

The following self-assessment assists the reader in identifying strengths and opportunities for improvement against the key performance dimension that we consider critical for succeeding with strategy management in the digital age.

For each question, any degree of agreement to the statement closer to one represents a significant opportunity for improvement (Table 4.1).

Table 4.1 Self-assessment checklist

Please tick the number that is the closest to the statement with which you agree 7 6 5 4 3 2 1	
My organization has a well-developed Strategy Map or similar	My organization has not developed a Strategy Map or similar
We articulate customer-facing strategic objectives from the starting point of the value they seek	We articulate customer-facing strategic objectives from a standpoint of what we want from the customer
Our strategic objectives have well defined descriptions	Our strategic objectives have poorly defined descriptions
We have a very good understanding of the causal relationships between objectives	We have a very poor understanding of the causal relationships between objectives
Generally, our people-related objectives are specific to the needs of the organization	Generally, our people-related objectives could be applied to any organization
In my organization, we very closely consider the relationship between intangible measures	In my organization, we only consider intangible measures in isolation
We place very high importance to the weighting of individual objectives	We place very low importance to the weighting of individual objectives
We have a very good understanding of the key strategic risks to the organization	We have a very good understanding of the key strategic risks to the organization

References

1. James Creelman, *Reinventing Planning and Budgeting for the Adaptive Enterprise,* Business Intelligence, 2006.
2. Mark L. Frigo and Richard J. Anderson, *What Is Strategic Risk Management?* Strategic Finance, April 2011.
3. Andrew Smart, James Creelman, *Risk-Based Performance. Management: Integrating Strategy and Risk Management,* Palgrave Macmillan, 2013.

5

How to Build an Agile and Adaptive Balanced Scorecard

Introduction

Being brutally honest, most Balanced Scorecards that we look at (by which we mean the scorecard of Key Performance Indicators—KPIs—targets and initiatives that support the Strategy Map) are of only marginal benefit to the organization (Fig. 5.1).

Too Many KPIs

Too often, Balanced Scorecards are excessively big with too many KPIs (and that have often been poorly designed) and the scorecard process is viewed by most people in the organization as little more than a quarterly chore of reporting a "score."

The strategy office (perhaps called an Office of Strategy Management—OSM—see Chap. 9: *Unleashing the Power of Analytics for Strategic Learning and Adapting*) spends all its time managing the scorecard, rather than using the scorecard to manage the business better. As we've been told on many occasions, "They're a bunch of data chasers." The process is fixed, inflexible, and cumbersome—the antithesis of agile or even adaptive.

As a starter, scorecards should have a limited number of KPIs (perhaps two for each of say 15 objectives), supported by about a dozen high-impact initiatives. Although quarterly reporting might still be appropriate (for formal governance reasons—although the reports should be much smaller than is typically the case), the data and insights must be available and applied on a more frequent basis (see also Chap. 9). Agility becomes possible.

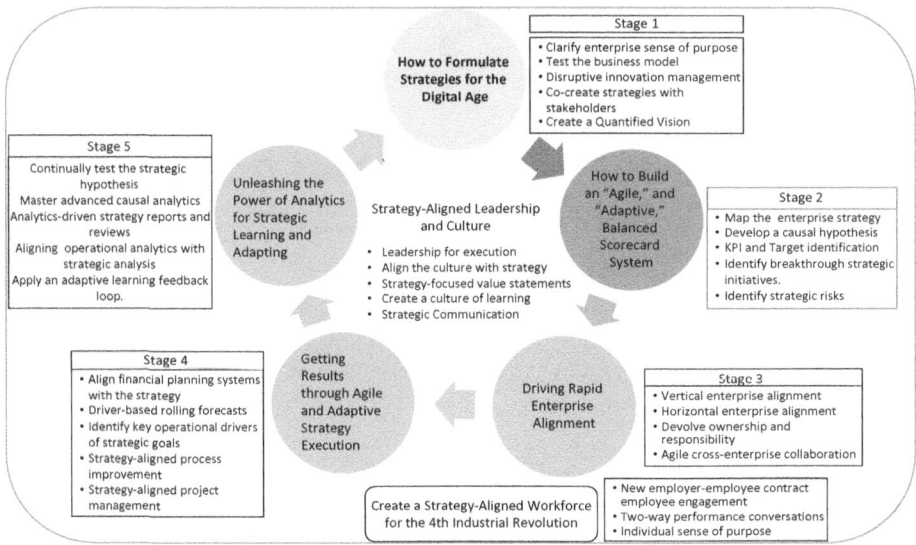

Fig. 5.1 Stage 2: How to build an "agile," and "adaptive," Balanced Scorecard System

Agility in Strategy Execution

Agility in strategy execution means being able to rapidly shift positions to exploit opportunities or mitigate risks—to capture, interpret, and act on data and insights in as close to real time as possible. Although strategy and operations are different things, this is where we are beginning to witness a blurring of the lines (see Stage 4, *Getting Results through Agile Strategy Execution*, Chaps. 7 and 8).

Before we get to execution and with the Strategy Map and objective statements agreed, we first need to identify the KPIs, set targets, and put in place the strategic initiatives. Alas, many organizations struggle with each of these three steps.

The Purpose of KPIs

We agree with the observation of Mihai Ionescu, Senior Strategy Consultant at the Romania-based Strategsys, on the challenges of identifying the relevant measures for each strategic objective. "This is where the operational and strategic views intersect," he says. You can often hear during workshops opinions such as, "We've always measured this KPI, so we should find a place for it in the scorecard," or "I'd like to see these additional KPIs in the scorecard, to have a more complete picture."

"This is what "translation" means for too many people; so, it's no wonder that the Balanced Scorecard soon becomes an unbalanced KPI system with lots of analytics that often succeed in masking its fundamental focus on the most important aspects and changes that will bring the strategy to life."

We also concur with the view of James Coffey, Principal of the US-based Beyond Scorecard, that the right strategic measures focus attention onto two things: (1) strategic discussions and (2) decision making.

Strategic discussions answer several key questions:

- Why did this happen?
- What do we need to do?
- What has changed that may require adapting our strategy?

This results in strategic decisions:

- What new initiatives do we need to improve performance?
- Is this a temporary dip with no action needed?
- What changes are needed to our strategy?

Coffey continues, "For strategic measures to be meaningful, leaders have to encourage accurate and honest reporting. The first time someone is punished for reporting poor results everyone else will do whatever is necessary to game the results."

He adds that developing measures requires addressing three concerns:

1. Will leadership care about the results? If all they do is review them but not discuss them, then they are not strategic.
2. Will they drive the right behaviours? If they are easily gamed, then it is vital to determine if the organization is doing the right things or merely doing what it takes to be green regardless of the overall impact on the strategic goals.
3. What are the risks associated with the performance? Strategic measures also drive risk discussions. They support assessing and understanding the consequences of not addressing risk.

Four Steps of KPI Selection

Driving the right behaviours is a critical consideration that we explain later in this chapter, (with the importance of strategic risks looked at in Panel 1). However, we will first outline the process for sequencing from objectives to KPIs. This comprises four steps.

1. Write the objective statement (which comprises both the strategic reason for the objective as well as how it will be achieved, see the previous chapter)
2. Identify the objective value drivers (the most critical capabilities and relationships that must be mastered if the objective is to be delivered).
3. Ask a Key Performance Question (KPQ) for each of the drivers (what is the most important information that we need to know about performance with regard to this driver).
4. Select the most impactful KPIs (and keeping in mind the objectives/KPIs that it drives in the ascending perspective).

Key Performance Questions Explained

Before providing a practical example, perhaps a note of explanation is required for KPQs, with which readers will likely be less familiar than KPIs.

Albert Einstein once said, "If I had an hour to solve a problem and my life depended on the solution, I would spend the first 55 minutes determining the proper question to ask… for once I know the proper question, I could solve the problem in less than five minutes."

The great physicist's words are always front-of-mind when we begin the process of selecting KPIs to support strategic objectives. Rather than which indicator to choose, better to think about which performance question any KPI will help to answer.

An innovation of LinkedIn influencer Bernard Marr (and explained fully in the book by Marr and Creelman *Doing More with Less: Measuring, Analyzing and Improving Performance in the Government and Not-For-Profit Sectors*, – [1],) a KPQ focuses on and highlights what the organization needs to know in terms of executing strategic objectives. Enabling a full and focused discussion on how well the organization is delivering on these objectives, KPQs also serve as an important bridge between organizational goals and KPIs. Too often, organizations jump straight from objectives to KPIs without truly understanding the performance issues the indicator will help address.

Case Illustration: Hospital Complex

As a practical example of the four steps and continuing with the hospital complex example from the previous chapter, the organization has an internal process objective to "assure service excellence & optimize the customer experience."

Step 1: The Objective Description

> Key to our success is ensuring that from entering to leaving the hospital the patient's experience is as comfortable and stress-free as possible.
>
> We will achieve this through implementing an end-to-end process that seamlessly integrates the process steps from the point at which the patient enters the hospital system, through the coordination of care whilst the patient is within the hospital, and finally to how the patient is discharged from the facility.

Step 2: Value Drivers

At the internal process and learning and growth levels, a driver-based model is applied to the achievement section (how the objective will be achieved). At the outcome level, it is applied to the results.

When using a driver-based model, the leadership team discusses what are *the* most important capabilities or relationships required for successful achievement. Perhaps three key drivers are chosen and are often listed in the "how," section of the objective statement. Here, the focus is on the wording of the objective statement, not the objective itself.

For the hospital complex, the key drivers identified were patient access, care coordination, and patient discharge (Fig. 5.2).

Step 3: KPQs

A KPQ for patient access might be, "how well do we understand the patient's needs from booking the appointment to entering the hospital?" For care coordination, "to what extent are departments sharing information to maximize the quality of care?"

Note that the KPQs are phrased to focus on the present and the future, not on the past. This opens a dialogue that enables managers to do something about the future rather than simply discussing past performance.

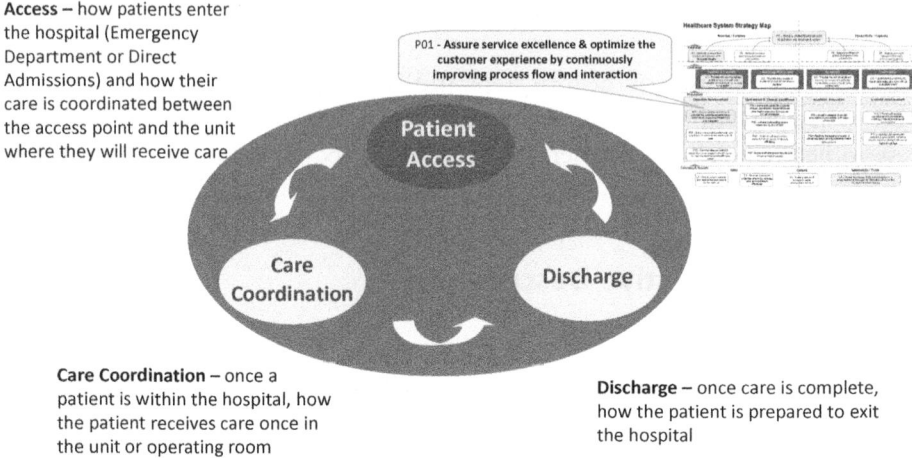

Access – how patients enter the hospital (Emergency Department or Direct Admissions) and how their care is coordinated between the access point and the unit where they will receive care

Care Coordination – once a patient is within the hospital, how the patient receives care once in the unit or operating room

Discharge – once care is complete, how the patient is prepared to exit the hospital

Fig. 5.2 Three sub-processes of hospital example strategic objective

Step 4: KPIs

The discussion around the KPQ invariable leads to the identification of the most relevant and impactful KPIs.

For the hospital complex, a KPI for, "how well do we understand the patient's needs from booking the appointment to entering the hospital?" might be: "percentage of patients that are very satisfied with the pre-admission process?"

For, "to what extent are departments sharing information to maximize the quality of care?" a KPI might be, "internal departmental satisfaction with the quality of patient information received from other departments?"

Case Illustration 2: Durham Constabulary

KPQs helped the UK-based Durham Constabulary convert the strategic objective, "Tackle Criminality," into relevant KPIs. The objective had three supporting KPQs, including "How well do we prevent people from becoming criminals?" which was assisted in being answered through a KPI that measured the "Number of first time entrants as a percentage of all persons arrested" as well as KPIs around re-offending rates and the percentage of the population who are offenders.

In *Doing More with Less*, Durham Constabulary's Head of Performance and Analysis, Gillian Porter commented that KPIs were still vitally important within the constabulary, but that the KPQs put these metrics into context. "…there's a tendency to measure things that are easy to collect data against and performance can get skewed as a result." She added that indicators can be dangerous things, especially when you get hung up on targets. "Now the focus is moving to identifying solutions rather than just performance against the indicators. It's becoming more about the conversation than the numbers" [2].

Interestingly, the Constabulary ensured that all strategic actions had to be justified against the objectives and KPQs: *note* - not the KPIs. Moreover, note the highlighting of the importance of the conversation. This signals that a KPI is the beginning of a conversation, not the end. The intent should be to improve performance, not simply to hit a target.

> **Advice Snippet**
>
> As a guide for crafting Key Performance Questions:
>
> 1. Design between one and three KPQs for each strategic objective on your Strategy Map
>
> Once you have clarified your strategic objectives and captured them within a Strategy Map, start designing KPQs. Between one and three KPQs for each strategic objective will suffice.
>
> 2. Create short and clear KPQs
>
> A good KPQ is relatively short, clear, and unambiguous. It should only contain one question.
>
> 3. KPQs should be open questions
>
> Closed questions such as "have we met our budget?" can be answered by a simple "yes" or "no." However, if we ask an open question such as "how well are we managing our budget?" the question triggers a wider search for answers and seeks more than a "yes" or "no" response.
>
> 4. KPQs should focus on the present and future
>
> Questions should be phrased in a way that addresses the present or future: "To what extent are we increasing our market share?" instead of questions that point to the past: "Has our market share increased?"
>
> 5. Use your KPQs to design relevant and meaningful performance indicators
>
> KPQs enable organizational leaders to identify the best data and management information they need to collect to help answer the key performance questions and therefore properly assess the progress of a strategic objective.

> 6. Use KPQs to refine and challenge existing performance indicators
>
> KPQs are useful for challenging and refining any existing performance indicators. Linking them to your KPIs can allow you to put them into context and justify their relevance.
>
> 7. Use KPQs to report, communicate, and review performance
>
> In performance reporting and communications, organizations should always put the KPQs with the performance data. This way, the person who looks at the data understands the purpose of collecting this data and is therefore able to put it into context.

The Balanced Scorecard Is Not a Measurement System

The four-step process is a relatively straightforward yet effective mechanism for identifying KPIs. Indeed, get the first three steps right and the KPIs are obvious, as numerous practitioners have told us.

Yet (as well as the criticality of the conversation) something we continually struggle to get organizations to understand is that, despite popular belief, the Balanced Scorecard is not primarily about measurement. As Mihai Ionescu, Senior Strategy Consultant at the Romania-based Strategsys says, "The Balanced Scorecard is not a measurement system. It is a strategic communication and decision-support system."

He continues, "It organizes, in a traceable way, the chains of hypothesis used in formulating the Strategy and in planning its execution, around the strategic causality concept, enabling the management team and the entire organization to participate in defining them, in identifying the invalid hypothesis, along the execution cycle, and in adapting the Strategy and the Strategic Plan accordingly."

KPIs are an important component of that system, but only third in the hierarchy of importance. First is the Strategy Map (that visually describes the intended results (outcomes) and the capabilities and relationships required for their delivery (enablers). Second is the initiatives (supported by process improvements) that deliver change. KPIs are simply a mechanism for monitoring performance to an objective.

No "Perfect" KPI

With the obsession on measures, there's a tendency to spend an inordinate amount of time finding the perfect KPI. As if by doing so, performance will automatically improve. Marginally perhaps, as people do pay for attention to what they are measured on (which is not always a good thing, see below) but step-change transformational change will certainly not happen.

Moreover, identifying the "ideal" KPI is notoriously challenging. We often end up selecting many KPIs, in the hope that, somehow, they will meld together into something that is perfect. Consequently, the scorecard system becomes a bloated and impermeable mechanism for capturing and reporting measures.

Quite simply, there is no such thing as a perfect KPI. As Bjarte Bogsnes, Senior Advisor, Performance Framework, at the oil and gas giant Statoil comments, "There are good KPIs and good combinations of KPIs, but not a perfect KPI. Believe me; I spent many years searching for them."

Moreover, although we have more than 500 years of experience of working with and evolving financial KPIs (double entry bookkeeping was invented by a Venetian monk in 1492), we have only relatively recently began working with non-financial measures. Therefore, our understanding is a lot less mature—thusly, a long way from anything close to perfection.

Also, seeking the perfect strategic KPI for a non-financial objective is something of a red herring. These KPIs differ from most of their financial counterparts in that they are rarely of value in isolation and are very difficult to benchmark across industries. Customer, process, and learning and growth KPIs work together, in a difficult-to-capture dynamic to deliver ultimate financial value (as we explained in earlier chapters).

Indeed, and particularly from the learning and growth perspective, organizations might have a KPI that is at best a proxy—in that it provides some, but by no means perfect, measure of progress. That is acceptable as these can be improved over time. This helps overcome the barrier of spending a lot of time worrying over what is the best KPI for creating a collaborative culture, for instance, and focusing on what needs to change (technologically and culturally) to enable collaboration.

As a result, the chosen KPI is still an indicator of progress to an objective, with an identified performance gap. But note the words of Bogsnes: "The I in KPI is there for a good reason, it is an indicator not an absolute measure of performance."

De-emphasize KPIs

Bogsnes argues that organizations should de-emphasize KPIs and emphasize actions. "Spend less time on KPI selection and expend more energy on thinking about the actions."

The marketplace moves way too fast these days to rely heavily on static KPIs and in spending months finding the perfect performance measure. The key to successfully implementing a strategy is speed and agility as well as adaptiveness. An overblown, over-engineered KPI system is a showstopper. To expand on a quote from John Ruskin. "In the final analysis, our goals and measures are of little consequence. Neither are what we think or believe. The only thing of consequence is what we do" [3].

We discuss "what we do" in Chap. 8: *Developing strategy-aligned project management capabilities* and the selection of initiatives below. However, before leaving the conversation on KPIs, there are two important topics to consider:

1. KPIs can be very misleading
2. KPIs can drive the wrong behaviours.

Misunderstanding here can lead to potentially catastrophic consequences and is often rooted in a lack of knowledge in the organization regarding how measures work.

The Science of Measurement

When we work with organizations to build scorecards, we always stress that those that work with measures need to understand at least the basics of how measures work. In assignments, we often have to explain the importance of understanding confidence levels and intervals when using surveys. This is not rocket science.

With at least the basics understood, we then progress to the basics of analytics. In time, they can mature to a more advanced understanding of measurement and analytics. But simply understanding the basics means that organizations do not spend endless amount of times collecting KPIs and then providing commentary that is at best of limited value or (not uncommon) downright dangerous as the so-called analysis leads to strategic, and often expensive, improvement interventions that are not addressing the problem—and perhaps exacerbating it.

It is a continued mystery how organizations are, as is typically the case, obsessed with measurement, but do not invest the time and money into teaching those that work with measures even the basics of the underpinning science. We would expect a finance professional to understand finance and the same for an IT specialist, but not for those working with KPIs. We need to redress this odd, and dangerous, omission.

The Dangers of Aggregation

Of the many common mistakes that organizations routinely make, consider aggregation.

Here's a task: put one leg in a bucket of boiling water and the other in a bucket of freezing water. On average, it's the perfect temperature. Herein lies a major issue with measurement that we commonly observe: believing that aggregated data is an insightful measure of performance. Of course, aggregation has some value as a high-level performance indicator, but without interrogating, the data beneath it can be very misleading.

Simpson's Paradox

As a powerful illustration, consider Simpson's Paradox: the paradox in probability and statistics, in which a trend appears in different groups of data but disappears or reverses when we combine these groups.

As a true example, a University in the USA was taken to court by a young woman that claimed gender bias on the basis that the annual admission data showed that significantly more boys were being admitted than girls. Sounds fair, yes?

However, the analysis of the data showed that generally girls were applying for the most competitive courses, whereas boys were more attracted to the less competitive courses. In reality, more girls were being admitted to both the more competitive and less competitive courses. Yet, when the numbers were aggregated, there were more boys admitted. Simpson's Paradox.

Consequently, whenever we are shown aggregated data (which on most scorecards are colour-coded), we always ask "but what does this mean?" Typically, the answer we get is "It shows we are performing well"—if it's green. Maybe it does or maybe it doesn't. We have no idea without looking at the underlying data. The downsides of colour coding are explored in Chap. 9: *Unleashing the Power of Analytics for Strategic Learning and Adapting.*

Herein lies another major issue with measurement: the belief that the reported KPI score is sufficient information for decision-making purposes. It is not. The top-level KPI "number" does not provide the full picture of performance. Back to the word "indicator."

Trend Analysis

A further area of concern is a failure to understand trends properly. It is the trend that is important, not the direct comparison between two adjacent numbers. This enables, as one example, performance to be analysed so to understand if any variations to performance can be attributed to reasons such as normal seasonal change or is due to negative influences that to redress require targeted interventions. As one practitioner noted, "We can make informed choices based on analysis that tell us that yes performance to a target has fallen by 10% but it's not significant so there's no need to do anything about it."

Moreover, good trend analysis might show that although a financial KPI is still green, it is trending downward, whereas a red KPI is trending upward and is forecasted to move into yellow soon. Much better to intervene to stop the downward trends than the one trending upwards. This points to one of the shortcoming of exception-based reporting, where all the conversations are about "apparently" underperforming KPIs.

Driving "Rational" Behaviours

A poor understanding of the science of measurement also means that organizations overlook the fact that measurement does not always drive the expected behaviours.

An Amusing Story

Here's an amusing true story. A few years back, the city of Mumbai had an issue with rat infestation. The rat population was growing rapidly and, of course, causing all manner of health concerns. Then a government official had a great idea. Mumbai also has a lot of poor people, so simply pay people to kill rats—deliver X numbers of dead rats and receive X rupees…brilliant!

It worked beautifully. The rat population declined rapidly. Nevertheless, after a few months it started to rise again, and the number of dead rats delivered for payment rose as well. No-one could understand why. Investigations found the answer—people were breeding rats. Makes sense as rats breed quickly, and they were proving a useful source of income to struggling folk.

A Tragic Story

The next true story is not amusing. A pizza company had an objective "Deliver Superior Customer Experience" and a supporting measure of "95% of Pizzas delivered within 15 minutes." This was based on research that found that customers wanted their food hot and delivered quickly. Makes sense, and to encourage the delivery of this customer experience, outlet managers' bonuses were based largely on this KPI.

One day, an outlet had an issue with its ovens and there was a panic about not hitting the target. The manager told a delivery boy to get on his motorcycle, hurry up, and get to the customers' homes quicker than he normally would have so as to hit the target. The young driver drove quickly, crashed, and died.

Be Careful What You Ask For

Both these amusing and tragic stories deliver the same message. Be careful what you ask for when creating KPIs and setting targets. They might just encourage "rational" behaviours that could be either positive or negative.

Dysfunctional behaviours (which are simply "rational" responses—that is, doing what is required to hit the target) triggered by a KPI are far from uncommon. It is well known (and we have seen this) for a manufacturing plant to bet set a target to reduce reported injuries, and for the target to be reached simply by only reporting serious injuries (that cannot be hidden). Performance does not improve, but the target is met. Put another way, "the target is hit, but the point is missed." There are many similar examples from all sectors, industries, and functions.

Identifying Rational Behaviours

When we work with organizations to select KPIs and targets, a step we always include is to get people to think about the behaviours the KPI might drive. We simply ask them to brainstorm and write down all the positive behaviours that might be encouraged and then the negatives. When done, we then discuss how to best encourage the former and mitigate the latter. Sometimes, the risk of dysfunctional behaviour is so great that the KPI must be rethought or abandoned: a simple exercise that can deliver a lot of benefits and save a lot of heartache.

Sometimes negative behaviours can happen as the KPI transitions from the design phase to reporting. We worked with one government organization that had set a KPI for the finance department of "90% of invoices paid within two months." This was something of a stretch for a government entity in that country. So, when we reviewed the finance scorecard, we were surprised to find the colour was green. It was being hit. This made little sense as in conversations with suppliers, a common gripe was that it took up to eight months to be paid. Clearly, something was amiss.

Auditing the KPI found that although the original intent was payment within two months of receipt, finance had changed this to two months from final sign-off, which—in this very bureaucratic organization—took six months. Again, performance did not change but the target was hit. No need for exception reporting here.

So, when designing KPIs and targets, think about the rational behaviours (positive and negative) that might be encouraged and plan accordingly. Also, ensure that the original performance-enhancing intent of the measure is not changed (oftentimes surreptitiously) during implementation. Indeed, a regular audit of the Balanced Scorecard is good practice (and when managers resist this, it is a strong indication that something just ain't right).

In addition, be particularly careful when bonuses are linked to KPI target achievement. An old adage says, "What gets measured gets done. What gets rewarded gets repeated." Be careful you don't simply end up rewarding more rats.

Advice Snippet

Organizations make a number of mistakes when working with KPIs. Amongst the most common are:

- A failure to understand the potential dangers of aggregating data. As well as hiding potentially damaging performance trends (hidden in the measures that are aggregated), they also might give a totally misleading view of performance: Simpson's Paradox.
- Not taking confidence levels and intervals into account leads to wasting time discussing statistically meaningless data. Best practice is to be 95% confident that the figure provide is correct to an error rate of two percentage points.
- Simply comparing one data point with the one previous. This provides a performance snapshot and is only meaningful when the organization has

> perhaps four or more time-based data points. For this reason, annual KPIs are of limited value when used over the timeline of a strategy.
> - Believing that the high-level KPI score is sufficient for analysis and reporting—it generally is not—and rigorous analysis of the data that underpins the KPI is important. Moreover, the I in KPI means indicator—not an absolute measure of performance.
>
> Underpinning this is a failure to provide staff that regularly collect and analyse data basic training in the metrology—the science of measurement.

Setting Targets

In his LinkedIn Blog *The Work is not the Problem*, Jeremy Cox, a Senior Consultant with Vanguard Consulting, neatly explained how the dysfunctional relationship between target-setting and the hierarchical (we would say Taloyesque) top-down style of management leads to "rational" behaviour [4]. "Top-down performance management causes systemic sub-optimisation and demotivation, as "what do I do to get the sale (and meet my target)" invariably wins over "what is the right thing for me to do for the customer?""

Cox rightly says that, many (we would say most) organizations remain wedded to the idea of targets, but,

> unfortunately, we know that the use of targets in a hierarchical system only engages peoples' ingenuity in managing the numbers rather than improving their methods. People's attention turns to being seen to meet the targets—fulfilling the bureaucratic requirements of reporting that which they have become "accountable" for—at the expense of achieving the organization's purpose. In simple terms, all this effort constitutes and causes waste—inefficiency, poor service and, worst of all, low morale.

He continues, "When targets produce unintended consequences, as they always do, managers react to the symptoms by doing more (adding targets) or less (subtracting targets, de-coupling incentives from targets) of the wrong thing."

He explains that doing the "wrong thing righter," is not the same as doing the right thing. "Unfortunately, leaders who grasp the damage wrought on their organization's performance by targets are forced to confront a dilemma because doing something different at the level of work requires de-constructing the system of management that creates and underpins the target regime. Arriving at this insight is by no means guaranteed – it is impossible for most managers (and government ministers) to imagine an alternative because of the degree to which targets have become embedded in our collective mind-set."

He argues that we do not need to learn how "to do change better," but that we need to change the way that we approach change. "Instead of separating decision-making from work, managers must learn to both study 'how the work really works' and then lead improvement activity in situ, with emergent change based on knowledge."

Cox's views support a key message of this book—that the underlying approach to managing organizations is no longer fit-for-purpose (if it ever was). In the digital age, we must reconfigure how we structure organizations, how decisions are made, and how we learn. Only then will "formal" frameworks such as the Balanced Scorecard be able to optimize performance fully and with an engaged workforce (see also Chap. 12: *Ensuring Employee Sense of Purpose in the Digital Age*).

While we figure what the new configurations will look like, we still need to set targets, and they are critical element of the Balanced Scorecard System. As Professor Kaplan has said, target setting is the least well developed of all the scorecard components. Oftentimes, targets are simply plucked from the air, akin to what often happens with target setting for the annual budget (the shortcomings of which we discuss in Chap. 7: *Aligning the Financial and Operational Drivers of Strategic Success*.)

One of the more prevalent criticisms of the annual budgeting process is that setting precise financial targets in say August 2017 to the end of 2018 is nonsensical, especially in fast-moving markets. The same applies to target setting on the Balanced Scorecard, which might set targets over a five-year period. Basic statistics tells us that setting precise targets over such a timeframe is nearly impossible with any reasonable degree of confidence (even in stable markets).

Precision and the Quantified Vision

However, we have a quandary here. With a quantified vision (see Chap. 3: *Agile Strategy Setting*), we typically include a precise financial number—perhaps over both five- and two-year timeframes. From this, the leadership team can understand the gap from where the organization is now, and the number expressed in the vision and the required contributions from revenue growth and cost reduction. Good analytics can help identify present and emerging market opportunities and internal cost reduction programs and work out timings and therefore set milestone targets.

Analytics will also enable an understanding of, for example, which components of customer satisfaction correlate with loyalty, repurchasing, and so on, enabling more targered performance improvement interventions and by implication better use of scarce financial and human resources.

That said, with so many unknowns these days, we must take care in turning these into absolute targets. The further out the target, the more it should be considered a stake-in-the-ground—a guide if you will.

From our experience, greater confidence and rigour can be applied over shorter-term horizons, much in keeping with best practice driver-based quarterly rolling forecasts (see also Chap. 7) where accuracy in the forecast is significant for the first couple of quarters but increasingly less so as the quarters stretch out. As time progresses, the organization can assess where it is against the quantified vision and judge what needs to be done to reach that financial target or whether it now requires revision.

Note that if a quantified financial target is very stretching, not hitting it over the timeline of the strategy is not necessarily a failure. It is a success if the organization is still outperforming its competitors and significantly improving performance. Indeed, for the quantified vision and for targets to KPIs, it might be worth considering ranges rather than precise figures, in line with good practice forecasting and our understanding of confidence levels and intervals.

A Target Is Not a Forecast

However, there's an important twist to this recommendation. A target *is not* a forecast—they are different things and have different purposes.

A target is what the organization would like to achieve, if all goes their way and it will typically be stretching, in that it represents an improvement from the present performance. This holds true for Balanced Scorecard targets as much as for the budget.

Conversely, a forecast is an honest assessment of likely future performance based on the most current data and information. The forecast should tell you whether you are on track to meet targets: it is a "health check" against targets, confirming their accuracy or providing an early warning signal of problems ahead.

Not appreciating the differences leads to significant dysfunctional behaviours—be that for budgeting or strategy. Applied to the Balanced Scorecard system, one of the authors recently reviewed the scorecard of a telecommunications company where the Head o the Office of Strategy Management (OSM) proudly pointed to all the green colours. Yet they were underperforming their competitors.

The fact is that their scorecard targets were tied to bonuses so, again as with the budget, managers fought for the easiest possible targets. Why wouldn't they? The fact is these scorecard targets were, in reality, forecasts. As with the budget, a rational outcome of incentive-compensation being based on targeted performance, but equally, it might be the consequence of a toxic culture where it is "better to be dead than red."

Assumption Management

So, organizational leaders must understand that a target and a forecast are different things, and this acceptance leads to significant benefits, especially when "key assumptions" are used to compare the two numbers, which might be external, (e.g. a macroeconomic development, a main competitor's behaviour) or internal (e.g. productivity gains, product launch success).

When assumption management is used, the infrastructure is in place for a rich and material conversation about what can and needs to be done to close the performance gap on the KPIs (the gap between present and desired performance, which collectively close the value gap between present and desired performance to the quantified vision).

For instance, if the forecast suggests that a financial target will be missed by 10%, it then becomes possible for an informed and honest exploration of the key assumptions (supported of course by good analytics) and to reach a conclusion as to what is really happening in that market and the likely consequences on financial performance. From that basis, there can be an agreement on next steps. Initiatives or other actions might be launched to close any performance gaps, or it might be agreed that market conditions have changed materially and thus the target cannot be hit.

The same thinking applies to non-financial targets. Perhaps something has happened that will materially affect a targeted customer KPI, for instance. The conversation can focus on whether this is a controllable issue that can be resolved through a particular intervention or is not controllable so not hitting the target can be accepted.

Choosing Strategic Initiatives

Of course, moving from present to targeted performance requires "work to be done." As an allegory, if someone decides to lose weight, they can set a target—that's easy. However, just setting the target won't change anything, as there would need to be "initiatives" around eating healthier/less and exercising, as examples. The same principle holds for improving performance on a scorecard, which is through continuous process improvements (see Chap. 7) and through strategic initiatives (Chap. 8 considers how to implement and manage initiatives). Here, we consider selecting initiatives.

Research Evidence

2014 research by the Palladium Group found that only 9% of almost 1300 firms surveyed believe they have the capability to optimize their strategic initiatives fully [5]. Findings such as these indicate there are significant problems here to address throughout the initiative management process.

One key reason for such disappointing findings is that, and unlike the selection of strategic objectives or KPIs, choosing initiatives means committing what might be scarce financial and human resources. A much more complex and politically charged process than choosing objectives and KPIs.

In addition, there are often too many initiatives on a scorecard (as there are usually too many objectives and KPIs). About 12 should suffice, and these should be high impact and, wherever possible, impacting more than one objective at the enabler lever. Initiatives should not be identified at the outcome level, as these are the result of the work done below.

Oftentimes, there are too many initiatives because a significant percentage are not initiatives, but regular work. As Kaplan and Norton explain, initiatives are one-off discretionary efforts and so are not repeated or business as usual. For example, publishing the annual report or completing scheduled maintenance problems are not initiatives.

Steps to Prioritizing Initiatives

In our work with hundreds of organizations, we have observed that there is a significant challenge in prioritizing initiatives amongst the many possible candidates.

An Initiative Inventory

We recommend that as a first step, all current initiatives should be inventoried and mapped against the objectives on the Strategy Map. Those that do not support an objective will not move to the next stage of prioritization.

Interestingly, it is our experience that this exercise delivers myriad benefits. It enables the culling of projects that might well have been launched for a good reason but are no longer strategically relevant. Furthermore, it invariably unearths instances of where similar initiatives exist in different parts of the organization and so enables the culling of initiatives that are misaligned with the strategy of the organization.

Other times, it is found that several initiatives are underway, which are really components of one bigger initiative; this enables the pooling together and streamlining of these efforts. Overall, the inventorying and mapping process by itself typically leads to significant cost-savings.

Prioritization Model

Nevertheless, due to resource constraints, organizations cannot usually fund all of the initiatives that successfully map to the Strategy Map. Therefore, all initiatives require a robust analysis of strategic, financial, and other benefits as well as costs and risk. A useful model for doing this—used by a government entity in the United Arab Emirates—is shown in Fig. 5.3.

It is also worth noting that not all selected initiatives have to be implemented at the same time. Sequencing of initiatives is important. All initiatives are of equal value over the lifetime of the strategy, but high-impact initiatives

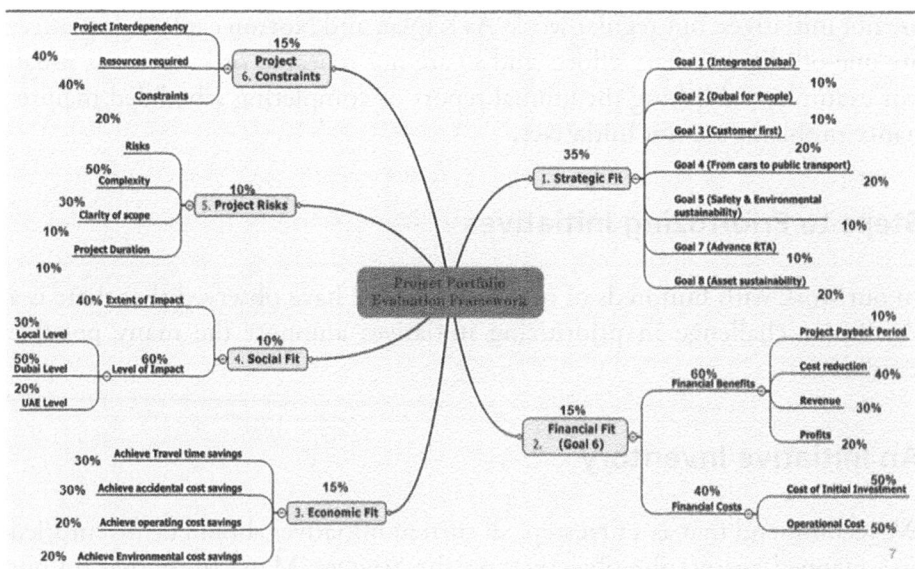

Fig. 5.3 Initiative scoring model

that drive quick wins might be prioritized. Moreover, there might be times during strategy implementation when one theme is more important than others and priority funding is directed to supporting initiatives (see also the previous chapter and the discussion on strategic themes).

Parting Words

With the organization-level Balanced Scorecard system established, the next step is to align each part of the enterprise to those strategic goals. Oftentimes, this is where any hoped-for performance agility gets smothered in overly restrictive diktats and controls, as well as ridiculously long timeframes for developing scorecard systems for lower-level units and functions. We explain how to overcome this agility-sapping conundrum in the next chapter.

Panel 1: Key Risk Indicators

Strategy cannot be managed effectively without understanding both the "performance" story (KPIs) and "risk" story (KRIs). The proper use of KRI provides for much greater insight into the future and promotes much greater quality of management conversation than can be gained by simply using KPIs.

To be fair, many organizations are using both KPIs and KRIs: although this is certainly an improvement on KPIs alone, they still tend to feed into different reporting and decision-making processes.

As the KPIs (which answer the question "Are we achieving our desired levels of performance?") and KRIs ("How is our risk profile changing and is it within our desired tolerance"?) are not integrated, they deliver a siloed and often competing, view of the organization and its performance. Therefore, the executive team does not have the appropriate data and information to inform the required high-quality management conversations that provide a more complete view of progress towards the strategic objectives or enable the trade-off between risk and reward to be discussed, understood, and acted upon.

A simple likelihood multiplied by impact equation is often used to assess the level of risk the organization is facing. KRIs provide a base of data and trend information that informs the calculation of risk exposures and informs management conversations as to current level of risk-taking, changes in risk-taking, and about how much risk needs to be taken to successfully deliver to the strategic objectives.

The other function of KRIs is that they help translate risk appetite into operational risk tolerances (expressed as thresholds around the indicators). If the organization has a high appetite, it would be expected that the threshold would be wider, allowing for greater levels of variation away from the baseline; whereas a low risk organization is going to have tight thresholds to promote a higher level of control.

> KRIs are typically derived from specific events or root causes, identified internally or externally, that can prevent achievement of performance goals. Examples can include items such as the introduction of a new product by a competitor, a strike at a supplier's plant, proposed changes in the regulatory environment, or input price changes.
>
> We strongly recommend the tracking of KPIs and KRIs on separate scorecards. A risk scorecard (or dashboard, to avoid terminology confusion) will complement the more conventional performance scorecard—but reported together.
>
> As Professor Robert Kaplan, in an interview with one of the authors in late 2015 on integrating strategy and risk management, said, "Identified risks should be managed through a separate risk dashboard. The scorecard is a about managing performance and the dashboard about managing risk, which are different things."
>
> For example, Infosys has a strategy focused on large contracts with large corporations. The concentration of revenues was identified as a significant strategic risk (a large account failure would show up on the income statement). The company identified a strategic risk indicator, credit default swap (CDS) rates, for its risk dashboard. If the CDS rate, the price for insuring against a client's default, went outside a specified range, then mitigation steps could be taken to cope with the client's increased risk.
>
> A risk mitigation might well be a strategic initiative that impacts both sets of indicators and, ultimately, the delivery of the strategic objectives.
>
> As we move deeper into the digital age (or the 4th industrial revolution), it will become increasingly critical to manage performance and risk equally as part of the strategy management process. As Professor Kaplan said:
>
> With good data and insights from both strategy and risk officers, the executive team can then make informed decision about how much risk they are willing to take in their strategy implementation efforts and how much to spend on strategy execution and risk management.
>
> With a deep knowledge of the performance/risk dynamic, managers might even take on more risk than their competitors—knowing that their risks are visible, that they are tracked through the strategic management system and that the limit of the risk taking is understood. In this way risk management becomes another tool for competitive advantage: as much about saying yes as saying no.

Self-Assessment Checklist

The following self-assessment assists the reader in identifying strengths and opportunities for improvement against the key performance dimension that we consider critical for succeeding with strategy management in the digital age.

For each question, any degree of agreement to the statement closer to one represents a significant opportunity for improvement (Table 5.1).

Table 5.1 Self-assessment checklist

Please tick the number that is the closest to the statement with which you agree		
	7 6 5 4 3 2 1	
When implementing strategy, my organization focuses on a small number of KPIs		When implementing strategy, my organization focuses on a large number of KPIs
We have a very good understanding of the purpose of KPIs		We have a very poor understanding of the purpose of KPIs
We use value drivers, or similar, in assigning KPIs to strategic objectives		We use brainstorming, or similar, in assigning KPIs to strategic objectives
My organization has a very good understanding of the strategic questions a KPI will answer		My organization has a very poor understanding of the strategic questions a KPI will answer
My organization places much more importance on initiatives/actions than on measures		My organization places much more importance on measures than on initiatives/actions
Those tasked with collecting and/or reporting KPI performance have a very good understanding of how measures work		Those tasked with collecting and/or reporting KPI performance have a very poor understanding of how measures work
When setting KPIs, we closely consider the potential negative behaviours that might be triggered		When setting KPIs we do not consider the potential negative behaviours that might be triggered
My organization has a very good understanding of how to set performance targets		My organization has a very poor understanding of how to set performance targets
We have an appropriate number of strategic initiatives		We have too many strategic initiatives
We have a very good process for prioritizing initiatives		We have a very poor process for prioritizing initiatives
We have a very good understanding of the difference between strategic initiatives and business as usual		We have a very poor understanding of the difference between strategic initiatives and business as usual
My organization has clearly identified the key risk indicators to track		My organization has not identified the key risk indicators to track

References

1. Bernard Marr, James Creelman, Doing More with Less: *measuring, analyzing performance in the government and not-for-profit sector*, Palgrave Macmillan, 2014.
2. Bernard Marr, James Creelman, Doing More with Less: *measuring, analyzing performance in the government and not-for-profit sector*, Palgrave Macmillan, 2014.
3. Adapted from a quote attributed to John Ruskin.
4. Jeremy Cox, *The Work is Not the Problem,* LinkedIn, June 2017.
5. James Creelman, Jade Evans, Caroline Lamaison, Matt Tice, *2014 Global State of Strategy and Leadership Survey Report*, Palladium Group, 2014.

6

Driving Rapid Enterprise Alignment

Introduction

In this book, we are focusing on strategic agility and adaptability—the need to be much more responsive in dealing with threats and opportunities. This is much more than speeding up the historically slow crafting of strategic plans and top-level Balanced Scorecard Systems. A much stronger and agile linkage between planning and execution is required, as is a more systematic strategic learning and action mechanism. We discuss these requirements within subsequent chapters (Fig. 6.1).

However, to align all of the organization to the strategy, then, we need an agile process and mechanism to align devolved objectives, Key Performance Indicators (KPIs), and initiatives/actions toward the strategy of the enterprise.

Traditional Approaches

Hoshin Kanri

Traditionally, many organizations have done this through approaches such as Hoshin Kanri, a seven-step process that involves the translation of a strategic vision plan into a set of clear objectives that are then realized through the execution of precisely defined tactical actions and projects.

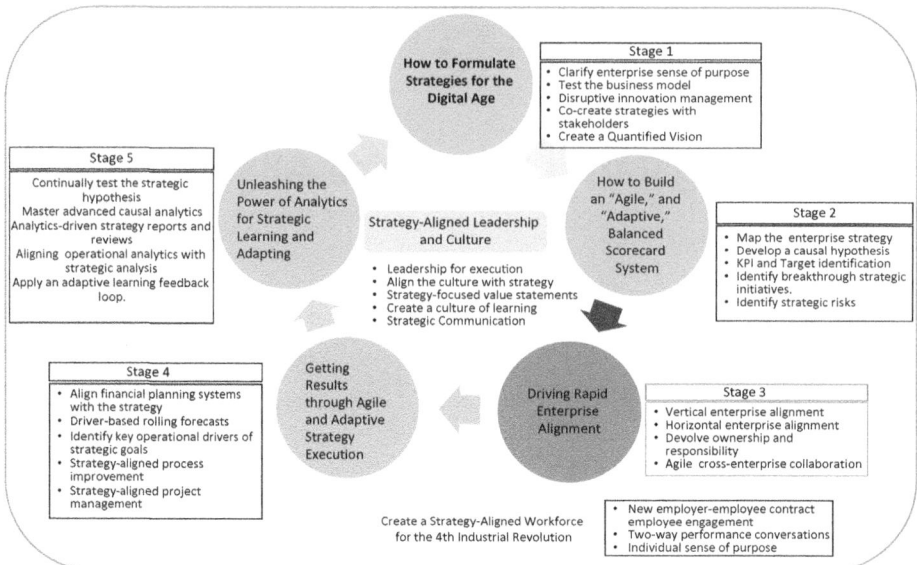

Fig. 6.1 Stage 3. Driving rapid enterprise alignment

Run to an annual cycle, Hoshin Kanri emerged in the 1960s in Japan out of the total quality movement and is based very much on the Shewhart/Deming plan-do-check act cycle.

Catchball

Hoshin uses a process called "catchball" to align all efforts across the organization. First, the catchball process translates strategies into increasingly lower-level objectives in a "cause-and-effect" way. Catchball then ensures that all the objectives at every level are well coordinated across process and functional lines.

A Strategy Map (in particular, at the theme level) can be used to guide the annual Hoshin Kanri process.

Thai Carbon Black Case Illustration

Palladium Hall of Fame inductee Thai Carbon Black—TCB—the largest single-plant producer of carbon black, the fine powder used in rubber manufacturing, chiefly for automobile tires, used both the Balanced Scorecard and Hoshin Kanri in a cascading strategy.

TCB's Quality Council (a senior executive team) decided to adopt the Balanced Scorecard, making it the central tool in its management arsenal. The Balanced Scorecard helped define and map strategy. Together with Hoshin-Kanri and Total Quality Management, it would help set goals and execute strategy. (Later, Six Sigma was used tactically within strategic initiatives).

The executive team sets a five-year strategic plan that it assesses and refines each year—along with the Strategy Map—in the Management Review and Quality Council meetings. The result is a one-year plan, known as the Presidents Policy, which contains key performance targets that align to each of the four Balanced Scorecard perspectives. Using catchball, TCB sets targets and supporting measures.

Senior managers set high-level strategic measures, targets, and "managing points" (key objectives, which must be approved by the president).

Through a dialogue with their direct reports, managers identify supporting measures for the level below, down to the supervisor level. "Checking points," components of the managing points, are in turn the responsibility of subordinate managers. At each level, action plans are launched to drive performance.

Balanced Scorecard

The corporate system serves as the steer for the divisional, strategic business unit and functional scorecards—or variations thereof. For more conventional Balanced Scorecard users, cascading (which, as we explain below, is *not* synonymous with alignment) is typically delivered through building Balanced Scorecard systems at a devolved level—a devolved approach.

Conventional Scorecard Shortcomings

Although widely used, and seemingly sensible for making "strategy everyone's everyday job," over the last couple of decades, practice has uncovered some potential shortcomings of the conventional scorecard cascading approach.

First, it is common for the cascade process to take an inordinate and enervating amount of time, especially when a large suite of scorecard systems is being built. By the time the final piece of the enterprise-wide scorecard jigsaw is in place, the picture the puzzle shows no longer accurately describes the strategy, or more usually, the required approach for its delivery. However, given the amount of time it takes to build, (sometimes up to a year—and even longer on occasions) there is little appetite to start again.

Secondly, the conventional cascade has always come with a suspicion of conforming to classical Tayloresque thinking. This well-crafted plan by leaders (as described in the top-level Strategy Map and Balanced Scorecard) is handed over "as is" (or close to) for departmental managers to implement.

The classic scorecard approach is to cascade themes, objectives, and KPIs according to three dimensions—identical, contributory, and unique.

Identical

Relevant elements are identical to those on the enterprise strategy map and scorecard.

Contributory

Some elements are translated to articulate the unique and direct contributions of the specific unit.

New

Unit develops new elements that describe indirect contributions to the enterprise scorecard.

To be fair, Kaplan and Norton recommend (and as taught in the Palladium Balanced Scorecard certification boot camp—which the authors of this book have delivered on numerous occasions) that cascaded identical elements (from themes to initiatives) should only reflect shared priorities—and that this is most often seen in the financial and learning and growth perspectives. Too often, this is not the case, and everything from the top level is cascaded. "Here's your objectives, KPIs, etc., – get on with it" (in keeping with the beloved Tayloresque diktat that managers think and reports do). After a while, the scorecard architects scratch their heads wondering why no one is taking this scorecard stuff seriously!

Too Cascade or Not

Before starting a cascade process, a useful question to ask is "do we need to cascade?" It is not always required, and usually not early in the scorecard journey. What precisely will cascading achieve? The stock answer of "aligning the organization to the strategy" is not enough, at least at the outset. Be clear about the tangible outcomes, in terms of results, working practices, structure, behaviours, and so on.

Also, consider whether the capabilities are in place to manage a suite of scorecard systems and how quickly the organization could change elements of the system if required.

There might also be times when a cascade is simply not required, which might be for specific organizational reasons, such as described in the Saatchi & Saatchi case study in Panel 1, where commonality and standardization across the globe (and in a very short timeframe) was a key reason for scorecard adoption.

Moreover, even if a cascade is required, it might be appropriate to pilot in a unit or function first, communicating progress and impact to the rest of the organization. During the pilot, the strategy team will also become more knowledgeable on what works, doesn't work, and likely resistance triggers. This might save a lot of pain later.

An Agile Approach

Case Illustration: Statoil

The Norway-headquartered oil and gas giant Statoil is a long-term user of the Balanced Scorecard system (with the first scorecard build in 1997). With about 700 scorecards in use worldwide by about 20,000 employees—from corporate to team level—the organization takes a very different approach to the classic cascade.

As a mature Balanced Scorecard user, Statoil (which calls it's system Ambition to Action) reached the conclusion that the conventional approach to cascading scorecards is not appropriate, at least not for them—a values-driven company. "When it comes to content, we prefer the word translation to cascading," explains Bjarte Bogsnes, Senior Advisor, Performance Framework. "Cascading is about corporate instructing that these are your objectives, KPIs, etc."

"However, in our culture we would very easily lose people's commitment, ownership and motivation if we controlled them in this way. Empowering teams to develop their own Ambition to Actions transmits a message that we trust them and believe in their abilities."

However, he does add that if such a translation should go wrong, then the team above should intervene, but this has never proven to be a real issue. "This is not a problem largely as a result of commitment to transparency – all Ambition to Actions are available for all to see online so inappropriate ideas are noted and not only by your manager. Transparency is a control mechanism for translation."

Bogsnes continues that used in the right way, the Balanced Scorecard is a great tool for supporting performance and helping teams to manage themselves. "Unfortunately, many scorecard implementations seem to be about reinforcing centralized command-and-control," he says. "Alignment is not about target numbers adding up on the decimal; it is about creating clarity about which mountain to climb."

Ambition to Action Explained

Ambition to Action has three purposes:

- Translate strategic choices into more concrete objectives, KPIs, and actions
- Secure flexibility and room to act and perform
- Activate Statoil's values and its people and leadership principles.

"Almost all our competitors have management systems that in some form aim to meet the first purpose, creating strategic goals," says Bogsnes, "but this loses what is key for success: autonomy and agility, trust and transparency, ownership and commitment. If your own Ambition to Action becomes nothing but a landing ground for instructions from above, both ownership and quality tend to walk out the door."

An Ambition to Action starts with an ambition statement, a higher purpose. "Call it a vision, call it a mission. We don't care, as long as it ignites and inspires," says Bogsnes.

The Statoil ambition is to be "Globally competitive – an exceptional place to perform and develop." This statement is translated into different versions across the company. "One of our technology teams, for instance, chose, "Execution for today, solutions for the future. In our team, we ...challenge traditional management thinking," explains Bogsnes.

"We try to connect and align all these through translation (each team translating relevant Ambition to Actions, typically the one above)," continues Bogsnes. "What should our Ambition to Action look like in order to support the Ambition to Action(s) above? What kind of objectives, KPIs and actions do we need? Can we use those above? Or do we need something sharper because we are one-step closer to the front line?"

"In short, we want Ambition to Action to be something that helps local teams to manage themselves and perform to their full potential, while we at the same time secure sufficient alignment."

He stressed that there are situations where instructions and cascading from above is necessary, "but this should be the exception and not the rule, which makes it more acceptable when it happens."

That said, there are occasional exceptions to this rule that might require more formal "instruction," at least over the shorter term, as explained in the Saatchi & Saatchi case study (Panel 1).

Change of Perspective Order

Interestingly, and very unusually, Statoil has switched the order of the scorecard perspectives. Finance is at the base and people and organization (its spin on learning and growth) on top. Bogsnes describes the thinking here. "We all know what happens in business review meetings when the agenda is tight and time is limited," he says. "Let's come back to people and organization next time, which then doesn't happen when next time comes. Those are not the signals to send if we claim to be a people-focused organization, so now People and Organization sits at the top. Another small gap closed between what we say and what we do."

Note that cause and effect still moves from people to finance, but the arrows point downwards and not upwards.

"We have a management model that is well suited to dealing with turbulence and rapid change. It enables us to act and reprioritize quickly so that we can fend off threats or seize opportunities," states Bogsnes.

As a powerful measure of how employees have responded to Statoil's translation rather than instruction approach, note that despite there being around 700 scorecards enterprise-wide, they are not mandated. "We have no detailed roll-out schedule. People are fed up with being on the receiving end of corporate 'roll-outs' again and again," says Bogsnes. "Instead, we focus on teams that invite us. Not once did we put our foot in the door because 'this is decided.' It takes longer, it looks messier, but change becomes real and sustainable."

We will return to Statoil in Chap. 11: *How to Ensure a Strategy-Aligned Culture*, where we explain how they have become a values-driven organization. More extensively, we discuss Statoil's unconventional approach to management in Chap. 7: *Aligning the Financial and Operational Drivers of Strategic Success*, where we explain how Statoil has become "budget-free"—it has not prepared an annual budget since 2006—and how it has introduced a more agile form of forecasting. Moreover, we explain how these approaches to financial management work alongside the Balanced Scorecard.

Key Lessons from Statoil

There is much to applaud and learn from Statoil's approach to cascading (and indeed its other managerial approaches). The conventional approach rarely leads to buy-in and is no longer fit-for-purpose in today's fast-moving markets. Spending up to a year designing a "beautiful" suite of aligned scorecards is simply pointless, and smothers strategic agility.

Rather, identify the critical (and very few) objectives and KPIs that must be devolved and then empower teams to build their own scorecard systems that describes what they want to achieve over the coming period.

Team Discussions

An important watch-out here often leads to frustration. The deeper into the organization the scorecard is cascaded, the less meaningful the term strategy becomes. We've lost count of the amount of times we've sat through (and even led) "Strategy awareness/alignment" sessions at departmental/team levels and witnessed eyes gradually glaze over and then people divert their gaze to that refuge of bored minds—the smartphone.

At departmental/team levels, we prefer to talk about the purpose of the group. We discuss the sense of purpose of the organization as encapsulated in the mission (see Chap. 2: *From Industrial to Digital-Age-Based Strategies*), then, how the group relates with other parts of the organization and their own sense of purpose: what they want to achieve over the short and medium terms and then why and how. See also Chap. 12: *Ensuring Employee Sense of Purpose in the Digital Age*.

At this level, focus is on the short and medium terms (no more than two years, so perhaps aligned to a mid-term quantified vision) and makes little more than a cursory mention of the longer term or where the organization wants to be in five years. Few of the staff will relate to this timescale as they are focused on day-to-day operations—and please do not mention shareholder value as people at lower levels have no real influence over this and will not be enthused about making shareholders richer.

Once the department/team's purpose is agreed, understood, aligned with that of the organization, and codified, it is important to quickly move to ensuring the interventions are in place (leadership support, HR, IT, etc.) to make it happen.

The sad fact is that no matter what is discussed in team meetings/focus groups, in most organizations, participants will believe nothing they say will be listened to and that nothing will change.

Bad Practice Example

One of the authors recalls doing a major project for an organization in which he ran focus groups in 18 departments (each of which had an enforced scorecard, which they took zero notice of). In *every* session, participants firmly stated that this was a pointless discussion because the senior leadership team never listened to them or did anything: but they did run similar exercises every two years or so, with a new and enthusiastic consultant.

With difficulty, they were encouraged to be forthcoming and provided lots of great (some amazing) ideas, which were subsequently submitted to the CEO. Two years later, the report was still sitting on his desk and no recommended interventions had been implemented. Still, the CEO raged about the failure of employees to contribute to the strategy. We guess he then ordered more focus groups.

Departmental Success Story

There's a slight twist to this tale. Although the 18 departmental scorecards gradually disappeared, along with the parent at the organizational level, one of the departmental heads was enthusiastic about the scorecard concept (having used it successfully in a previous company). He called one of the authors to help him develop a Strategy Map and Balanced Scorecard. This was completed in three one-hour sessions. Two years later, it was the only part of the organization with a scorecard and had been devolved to team level at the request of the teams (in line with the advice from Bogsnes). The manager of this department informed the author later that the CEO had said to him, "Why are you the only part of the organization that ever mentions strategy?"

As well as leading to greater buy-in and ownership, allowing departments, and the like, to build their own scorecard systems transmits the message stressed by Bogsnes that senior management trusts their employees and believes in their abilities. Proper governance still ensures alignment, but guided by flexibility and empowerment instead of rigid imposition.

Structure

Structurally, the process for building cascaded Balanced Scorecard Systems is the same as for the corporate or top level. Based on the scorecard above, build a Strategic Change Agenda for the devolved level, followed by objectives, KPIs, targets, and initiatives. This does not change, except that, and for alignment purposes, a critical input is the work done at the level above.

> **Advice Snippet**
>
> When cascading a Balanced Scorecard system, keep the following in mind:
>
> - Keep any mandated objectives, KPIs, and so on, to the critical few. For instance, for some organizations, this might relate to safety or environmental issues. If it's mandated, be very clear why this is the case.
> - Focus the conversations on translating higher-level objectives into what the cascaded unit/department and so on wishes to achieve.
> - The deeper into the organization the scorecard is cascaded, lessen use of the term strategy, perhaps replacing it with sense of purpose (although still communicating the organizational strategy).
> - Focus on the midterm rather than the longer term, as this is more relevant and triggers a greater sense of urgency.
> - Pay attention to the capabilities the unit, and the like wish to develop as this opens up developmental opportunities for staff, which strengthens the likelihood of buy-in.
> - Make sure that agreed interventions that are required by levels that are more senior actually happen.

Alignment and Synergies

As stated earlier, alignment and cascade are not the same thing—although cascading is a central plank. Whereas cascading implies a downward motion, alignment is more multi-directional, so horizontal or matrixed, or even across partnerships (we explore partnership alignment in Chap. 13: *Further Developments, Driving Sustainable Value through Collaborative Strategy Maps and Scorecards*). Alignment also encompasses capabilities, such as for human and information capital (see Chap. 12: *Ensuring Employee Sense of Purpose in the Digital Age*).

Furthermore, and as Ionescu comments, alignment is also about the strategic dialogue, "which allow the ideas, opinions and suggestions to bubble-up from the individual level to the departmental level and from here to the organizational level." Returning to theme of Chap. 4, *Strategy Mapping in Disruptive Times,* alignment might be more quantum mechanics than Newtonian.

Alignment: Using the Balanced Scorecard to Create Corporate Synergies

Of Kaplan and Norton's canon of five books, the one that has received the least attention is book four: *Alignment: Using the Balanced Scorecard to Create Corporate Synergies*, [1].

This is a shame, as here they provide a very useful framework for driving alignment and, from this, synergies across the enterprise, both vertically and horizontally. Particularly powerful is the enabling of synergies within even the most diversified organization.

Strategy Maps and Diversified Organizations

We will pause here to make an observation on developing Strategy Maps for diversified organizations. Over the years, we have received questions many times on how to do this, as it proves to be a difficult challenge. The answer is simple—don't.

A Strategy Map describes the value proposition to a specific set of customers, often served by a strategic business unit. In a diversified organization, there will be very different customers with very different value propositions, so the required objectives will also be diverse. In such cases, it is better to create separate scorecard systems. That said, and what was clever about Kaplan and Norton's recommendations was that this does not mean that scorecard thinking cannot be applied at the corporate level of a diversified organization. However, a Strategy Map is not deployed.

Enterprise Synergy Model

In the Kaplan and Norton model, the four scorecard dimensions remain the same, but the naming is changed. Rather than perspectives, we have synergies; so, for example, the "Financial Perspective" becomes "Financial Synergies."

In addition, the overriding questions that support conventional perspectives are different. Whereas for the classic scorecard we ask, "How do we create value in the eyes of shareholders/funders" here, the question is, "How can we increase the shareholder value of our Strategic Business Unit (SBU) portfolio?" For instance, creating synergy through effective management of internal capital and labour markets. The point here is that the enterprise level company is the funder of SBU operations, so it should look to allocate capital that drives collective value, perhaps by funding a shared services organization or common ERP system.

A more significant difference is for customer synergies. Here, the corporate centre, if you will, is not seeking to deliver a value proposition to a customer segment to secure revenues. The enterprise owns the SBUs—who obviously cannot choose to defect to a competitor. So here, the focus is on "How can we share the customer interface to increase total customer value?" Cross-selling is one way to do this.

For internal process synergies, the question is, "How can we manage SBU processes to achieve economies of scale or value chain integration?"- A shared services centre is one example. At learning and growth, we ask, "How can we develop and share our intangible assets?" so here, we look at areas such as common IT systems, leadership development programs, common values, and so on.

Driving Out Complexity

As much as anything, this model speaks to the importance of driving out complexity. Increasing layers of complexity accompany any organizational growth. Oftentimes, the original value proposition of the firm—which provided the success that enabled growth in the first place—gets diluted, and even lost, in the day-to-day battles of managing a larger organization. The sense of purpose is forgotten.

As an aside, one of the authors of this book once worked with a small, and very successful, organization that was about to set off on a trajectory of significant growth. The advice given was to frame the original Strategy Map and hang it on the CEOs wall so that the organization would always remember what made them great in the first place.

Why Simplicity Matters

Continued research by firms such as the US-headquartered The Hackett Group, with which one of the authors has enjoyed a two-decade-long working relationship, has repeatedly shown that the most effective way to reduce complexity is through standardizing, automating, and simplifying core processes. This makes it easier to adjust to all types of business change.

In the 2009 book, *The Finance Function: Achieving performance excellence in a global economy*, written by one of the authors, Jodiann Hobson, then Senior Business Advisor, The Hackett Group (now Principal Business Architect for the Australia-based Westpac Group), made this observation on how world-class finance organizations had managed to reduce costs significantly through driving out as much complexity as possible. Being world class (according to Hackett's assessment criteria) also meant they were simultaneously delivering greater value to their internal and, where appropriate, external customers. "World-class companies have managed to reduce finance cost (as a percentage of revenue) despite the increased complexity and volatility of their operating environment – this is through their laser focus on continuous improvement and investment in standardized processes and technology," she says [2]. Hackett research finds that the same principles of standardization, and so on, hold true for other support functions such as IT, HR, and procurement.

The purpose of including these observations here is this: the digital age will be increasingly global in nature and characterized by significant complexity—markets, customers, processes, employees, and technology. If organizations are to be agile and adaptive in strategy execution, they will need to ensure that this is not constrained by complexity: be that structure, policies, or fractured processes.

Whether diversified or not, organizations can use the thinking of the Kaplan and Norton Enterprise Synergy Model to drive greater value to customers, while still maintaining the required cost position. Much of this thinking might be applied to the initiatives chosen to appear on the corporate-level scorecard, which might then be cascaded to lower-level units. This might be particularly impactful within the learning and growth perspective.

Parting Words

Nevertheless, as Hackett research has uncovered, the key is to know when some level of complexity is still required for a competitive advantage. Alignment is essentially about getting collective focus on an agreed set of outcomes. Given the dynamic nature of markets, organizations, and people, this will always mean the acceptance of complexity and the dexterity to respond.

As with building the Strategy Map and Balanced Scorecard, when driving out complexity, take heed of the further words of Professor Albert Einstein, "Everything should be made as simple as possible, but not simpler."

> **Panel 1: Exception to the Rule – When Instruction Makes Sense. Saatchi & Saatchi Case Illustration**
>
> In the mid-1990s, the communications agency Saatchi & Saatchi Worldwide was on the brink of bankruptcy, and a new Chairman and a new CEO were recruited to drive an aggressive turnaround. A Strategy Map of just 12 objectives and a Balanced Scorecard of 25 KPIs was created for the corporate level.
>
> All 45 country-based business units had to work with this system (with some modifications in targets, to reflect local conditions) but scorecards were not created at any lower level. The 45 units rolled up to 11 regions, which then got consolidated to make the worldwide enterprise scorecard. Performance was typically discussed from a regional perspective.
>
> Now, the reason why Saatchi & Saatchi purposefully did not create a large suite of scorecards or even provide some degree of autonomy in unit-level scorecards was relatively simple.
>
> In the decade or so preceding their financial woes, the organization had grown through aggressive acquisitions across the globe. Each country-based acquisition was essentially allowed to continue as before, but under the Saatchi & Saatchi brand. As a result, there were basically 45 different organizations, with very different processes, policies, and cultures.
>
> When deploying the Balanced Scorecard System to drive transformation, the new executive leadership team decided to create a new structure that drove common practices, processes, and so on, enterprise-wide. Moreover, recognizing that many different cultures were operating within the organization, the people and culture perspective (its version of learning and growth) included just one objective, "One team: One dream: create a rewarding, stimulating environment where nothing is impossible." The goal being to forge a common culture.
>
> In short, the organization did not cascade, and so did not provide the autonomy for devolved units to build their own scorecard systems because there was a pressing need to drive discipline around commonality in processes and culture.
>
> For an organization that was on the brink of bankruptcy, this was a sensible choice. As we explain in Chap. 10: *How to Ensure a Strategy-Aligned Leadership*, different strategies often require different leadership styles and, sometimes, a period of "instruction" is required, but always over the shorter term. It is not sustainable over the longer term—especially when the organization is no longer in danger of extinction.

Self-Assessment Checklist

The following self-assessment assists the reader in identifying strengths and opportunities for improvement against the key performance dimension that we consider critical for succeeding with strategy management in the digital age.

For each question, any degree of agreement to the statement closer to one represents a significant opportunity for improvement.

Please tick the number that is the closest to the statement with which you agree		
	7　6　5　4　3　2　1	
My organization has a very good process for cascading strategic objectives		My organization has a very poor process for cascading strategic objectives
Strategic objectives are generally set by the individual units/teams		Strategic objectives are generally enforced in a top-down fashion
When setting team objectives, we closely consider the "sense of purpose" of the team—what they want to achieve		When setting team objectives, we do not consider the "sense of purpose" of the team—what they want to achieve
Generally, units/teams have fully bought in to their objectives/KPIs		Generally, units/teams have not bought in to their objectives/KPIs
We have very well-established processes for gaining synergies across the organization		We have very poor processes for gaining synergies across the organization

References

1. Robert S. Kaplan, David P. Norton, *Alignment: Using the Balanced Scorecard to Create Corporate Synergies,* Harvard Business School Press, 2006.
2. James Creelman, *The Finance Function: Achieving performance excellence in a global economy,* Optima Publications, 2009.

7

Aligning the Financial and Operational Drivers of Strategic Success

Introduction

In the next two chapters, we explain how to execute strategies. As explained in the previous chapter, the Balanced Scorecard system will not implement itself—it is simply the plan. Implementing strategy is through the delivery of the chosen initiatives, (covered in the next chapter) as well as through strategically aligned process improvements (which we deal with in this chapter). To identify process improvements, we must first align operations to strategy, which also comprises better linkage to financial plans (budgets being the financial component of the annual operating plan) (Fig. 7.1).

Aligning Budgeting with Strategy

In a 2001 interview, Professor Kaplan said, "One aspect of the Strategy-Focused Organization that has lagged is the integration with the budgeting system.... if we don't establish the link with budgeting, then scorecard initiatives may wither" [1].

More than 15 years later, the linkage is generally still weak. The long-established yet dysfunctional budgeting dance is still a ritualized annual event that most organizations feel compelled to honour (even though most see its stupidity). To Kaplan's point, with strategy being something that cannot be shoehorned into a financial year, the strategy dance takes place somewhere else and to a different rhythm. As Anders Olesen, Director of the Beyond

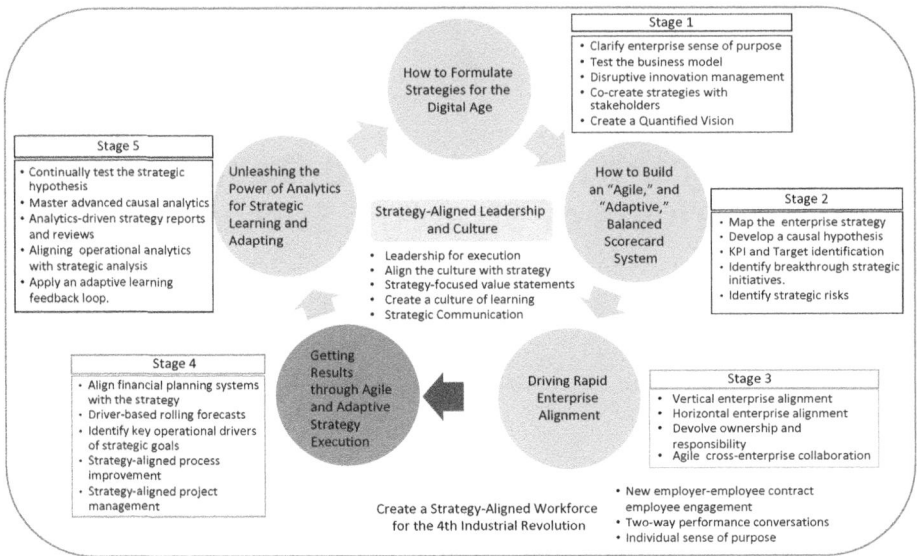

Fig. 7.1 Stage 4: getting results through agile and adaptive strategy execution

Budgeting Institute says, "The budget is supposed to be the detailed plan for year one in the strategy period. However, experience shows that very little of the budget work has anything to do with strategy."

The Shortcomings of Budgeting

The shortcomings of the annual budgeting process are well known and generally accepted. Most notably:

- Too lengthy (often starting in, say mid-2018 to set targets for the end of 2019—madness in many industries)
- Out-of-date the very day it is published
- Too detailed (with 100s of line items)
- Inflexible (locks resources in for a year)

Perhaps the most damaging shortcoming is that it is a process of target negotiations, largely due to the link with annual bonuses. A key movement of the annual dance is when corporate sets a target (knowing it will be rejected as it is set higher than required) and sends it to unit managers, who come back with a revised target (less than they know they can achieve) and, eventually, a compromise is reached. Jack Welch has called it an "exercise in minimization." In his seminal work *Winning*, he calls the budgeting process, "the most ineffective practice in management. It sucks the energy, time, fun and big dreams out of an organization…In fact when companies win, in most cases it is despite their budgets, not because of them."[2].

The Three Roles of Budgeting

A particular weakness of the budgeting process, and one that which many organizations are unaware of, is that budgets essentially serve three purposes at once—which are three different processes: target setting, forecasting, and resource allocation.

As Bjarte Bogsnes, Senior Advisor, Performance Framework, Statoil, and Chairman of the Beyond Budgeting institute comments, "The three purposes can't be meaningfully handled in one process resulting in one set of numbers. A target is what we want to happen. A forecast is what we think will happen, whether we like what we see or not. And resource allocation is about trying to use our resources in the most optimal and efficient way."

"An ambitious sales target can't at the same time also be an unbiased sales forecast," he continues. "And you rarely get a good cost forecast if the organization believes this is their one shot at access to resources for the next year."

Case Illustration: Statoil

Under the guidance of Bogsnes and with the full support of the executive committee, Statoil (a Norway-headquartered energy company that operates in over 30 countries, with $61 billion in revenues in 2017) has pioneered approaches to separate these out into three processes. Furthermore, they link firmly to the Balanced Scorecard System, (which it calls "Ambition to Action"). Moreover, Statoil does not produce an annual budget and hasn't done so since 2006 [5]. Figure 7.2 shows the organization's alternative approach to budgeting.

Within Statoil, Ambition to Action (which has provided the strategic direction for the enterprise since 1997) is the only mechanism used for target setting. This comprises a small number of objectives, KPIs, and so on.

Wherever possible, the organization (which has about 700 Ambitions to Action worldwide) uses relative, instead of absolute and decimal-loaded, targets. Bogsnes explains the value, "Relative targets address how we are doing compared to others, internally or externally, instead of a myopic focus on fixed and decimal-oriented numbers."

Statoil's main financial targets are set against a peer group of 15 other oil and gas companies and it aims to be in the first quartile on return on capital

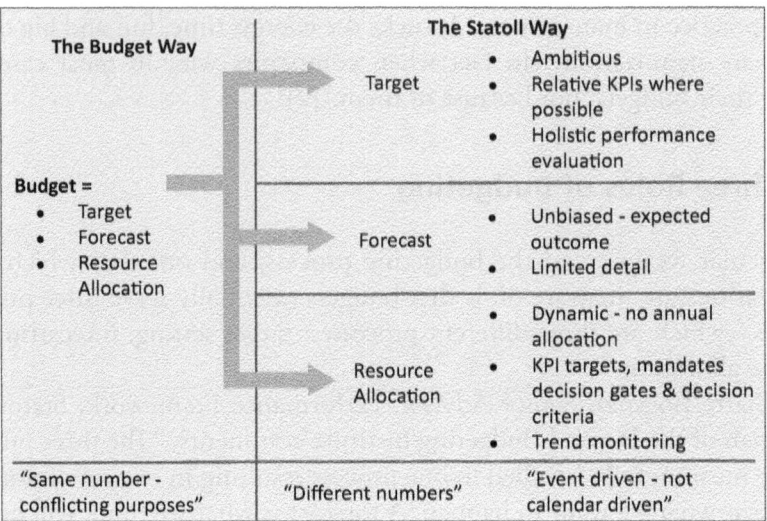

Fig. 7.2 Statoil's alternative to the budget

employed, above average on shareholder return, and in first quartile on unit production cost. "These are the kind of financial targets our Board approves," Bogsnes explains. "They do not approve a budget. The two first metrics are also key in our common bonus scheme. Everybody in the same boat; us against the competition."

With regard to target setting, he adds that he has yet to hear a team coming out low (given they find the benchmarking fair and relevant) announce that they have no ambitions about climbing on the ranking. "This is a much more self-regulating approach compared to the traditional budget game, which stimulates the very opposite mind-set, the one that drives managers to negotiate for the easiest achievable number."

As Statoil has taken out much of the gaming bias that typically accompanies target setting or resource allocation, the quality of its forecasting has also improved. A long-time user of rolling forecasts (explained below), the organization has introduced dynamic forecasting. "A rolling forecast is done on a fixed frequency and on a fixed time horizon across the company, often quarterly and five quarters ahead," says Bogsnes. "We want forecasts to be updated when something happens, and as far ahead as relevant for each unit." He explains that, in the energy business, different parts of the organization (from oil exploration to retail) work to very different rhythms and so "forcing standard forecasting horizons doesn't make much sense."

Also introduced has been dynamic resource allocation, which provides much bigger and more flexible decision authorities to local teams and to a much more dynamic rhythm, according to Bogsnes, who provides this analogy. "Imagine a bank informing its customers, 'We have now changed our hours, so if you want to borrow money, we are now only open in October.' It sounds ridiculous—but isn't this exactly what people in companies experience every year in the budget process?"

He explains that when making cost decisions, they do not want managers to ask, "Do I have a budget for this?" but

- Is this really necessary?
- What is good enough?
- How is this creating value?
- Is this within my execution framework?

"In addition, we must always consider capacity, both financial and human. As things look today, can we afford it, and do we have the people to do it? This information would typically come from our latest forecasts."

For operational or administrative cost, with fewer discrete decision points than projects have, Statoil offers a menu of alternative mechanisms for the business to manage its own costs. These include "burn rate" guidance ("operate within this approximate activity level"), unit cost targets ("you can spend more if you produce more"), benchmarked targets ("keep unit cost below our peers' average"), profit targets ("spend so that you maximize your bottom line"), or simply no target at all ("we'll monitor cost trends and intervene only if necessary") Fig. 7.3.

Last, but not least, Statoil completes its performance management framework by deploying a more holistic performance evaluation, with hindsight insights as a key component, which Bogsnes calls, "the wealth of information unavailable at target setting time."

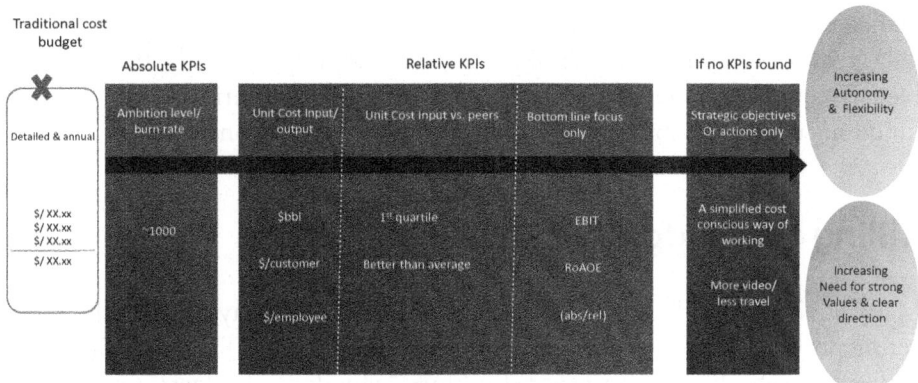

Fig. 7.3 Statoil's menu for managing costs

He continues, "We ask, for instance, have we really moved towards our longer-term objectives? Was there significant tailwind or headwind to take into account? Are results sustainable; will they stand the test of time?"

Bogsnes concludes that, "Combined with the Beyond Budgeting principles [which provide guidelines for alternative approaches to budgeting, (see advice snippet below)], Ambition to Action becomes a much more robust management model, solving many of the problems that can occur in conventional Balanced Scorecard implementations." Thus, addressing the issues outlined by Kaplan at the start of this chapter.

Agile Financial Management

An important lesson we learn from the Statoil example is that becoming more agile and adaptive in strategy execution requires significant alterations to other management models and processes, most notably, financial. If financial planning and resource allocation are not more "dynamic," then it is impossible for strategy to become so. Becoming agile/adaptive truly requires a fundamental overhaul of how the organization is managed strategically, financially, operationally, and in managing people. In short, finally breaking free of the straitjacket that is Taylorism.

Armen Mnatsakanyan, CEO of the Moscow, Russia-based management consultancy ConconFM puts this spin into the dysfunctional, silo-based structure of organizations: "This problem has no solution within the Finance Department! We need to look at the entire "Company Management System." Considered from the point of view of the "Food Chain of the Company" (the critical functions that deliver value and horizontally), we should build the work of the financial department not as a self-sufficient system within the company (as it is typically practiced), but as an element of the end-to-end management system of the whole company."

"Organizations have to rebuild (or rather adjust), not only the Finance Department, but other departments. This is the reformation of all the key functions under the control "of the Food Chain of the Company.""

Killing the Budget

Many advisors have long argued that the most obvious way to solve the budgeting problem is to stop doing it, as in the case of Statoil. Simply asking, and without the shackles of ingrained thinking, "why do we budget?" can lead to interesting insights.

Interestingly, over many years, research by firms such as the US-headquartered The Hackett Group have found that most organizations (including the leaders of the finance function) are very (oftentimes painfully) aware of the shortcomings of the budget, and while many plan to replace it with something more useful and agile, few actually do. The dance continues.

Oftentimes, this is because although dysfunctional, managers are well rehearsed in the moves of the dance (and many do it very well to their advantage) and there is an absence of a generally accepted alternative model. Organizations such as the Beyond Budgeting Institute have been working for several decades to provide other options, with Statoil being a leading example.

However, for many organizations, jettisoning the budget would be a fearful step. However, this in and of itself is not a showstopper to becoming more agile. Even with the annual budget left in place, organizations can become more agile by de-emphasizing the budget and emphasizing rolling forecasts.

Rolling Forecasts

Through this approach, a lighter budget (with fewer line items and completed more quickly than is generally the case) sets annual targets, while the rolling forecast becomes the main process for steering the organization on a day-to-day basis. The annual target retains its high level of importance, but senior managers are more interested in the veracity of the data that they receive on a regular basis from the rolling forecast, as this tells them what the future will look like and where course corrections are required (thus linking into the performance intervention element of target tracking).

A Rolling Forecast Explained

A rolling forecast is a projection into the future, based on past performances (trends) that are updated (typically on a quarterly basis, but perhaps monthly in fast-moving markets) to incorporate input and information reflecting changing market, industry, and/or business conditions.

At stated earlier and in the previous chapter, a forecast (be it rolling, dynamic, or, more conventionally, to the end of the financial year) is not a target, but rather, a best (and importantly honest) prediction as to the organization's financial and operational performance over a certain time horizon. For a rolling forecast, that time horizon can be 12, 18, 24, or any number of months or quarters ahead from the present. It "rolls" because as time moves forward, so does the time horizon of the forecast, unlike a traditional budget cycle that ends at a fixed point in time. Generally, there will be greater precision over shorter timespans than longer.

Unlike traditional budgeting, where an organization basically starts from scratch and marshals significant resources (and contends with ongoing negotiations), a rolling forecast involves only minor tweaking—as you continually update on a short-term basis. This saves time and resources.

Importantly, as the rolling forecast typically looks at least 12 months ahead, there is always a forecast to the end of the financial year, so enabling the assessment of the likely performance to the annual budget. A commonly held misunderstanding of the rolling forecast is that it replaces an annual plan/budget. This is not true; it is usually used alongside the annual plan/budget.

As the rolling forecast almost invariably looks out beyond the fiscal year, it provides useful insights into the robustness and continued appropriateness of the mid-term quantified vision and targets (see Chap. 3: *Agile Strategy Setting*), which in turn provides an early warning signal regarding the likelihood of delivering on the five-year strategic plan (Fig. 7.4). For this reason, we recommend integrating rolling forecasts into the strategy review meeting (see Chap. 9: *Unleashing the Power of Analytics for Strategic Learning and Adapting*.)

Constructing a Rolling Forecast

As with a de-emphasized annual budget, we recommend restricting the rolling forecast to a limited number of metrics. All that happens when an organization de-emphasizes the budget in favour of an overly detailed forecast is that one broken process is replaced by another, which will not lead to performance improvements. Indeed, research conducted by one of the authors of this book, a few years ago, into the forecasting practices of large organizations found that the forecast (typically to the end of the financial year) was little more than a mirror of the budget.

Moreover, we found forecasting to be very politically/culturally biased. If the forecasters knew that the CEO/CFO preferred upbeat forecasts, then that was what he/she got. Conversely, if they liked pessimistic forecasts, then that is what they got. Hardly a useful alternative.

Driver-based models are effective for constructing a rolling forecast, (Fig. 7.5). This entails determining the drivers of major costs, then considering them according to two variables: rate and cost. The payoff is twofold: adjusting a simpler budget is relatively easy and far more valuable conversations with internal customers.

Aligning the Financial and Operational Drivers of Strategic Success

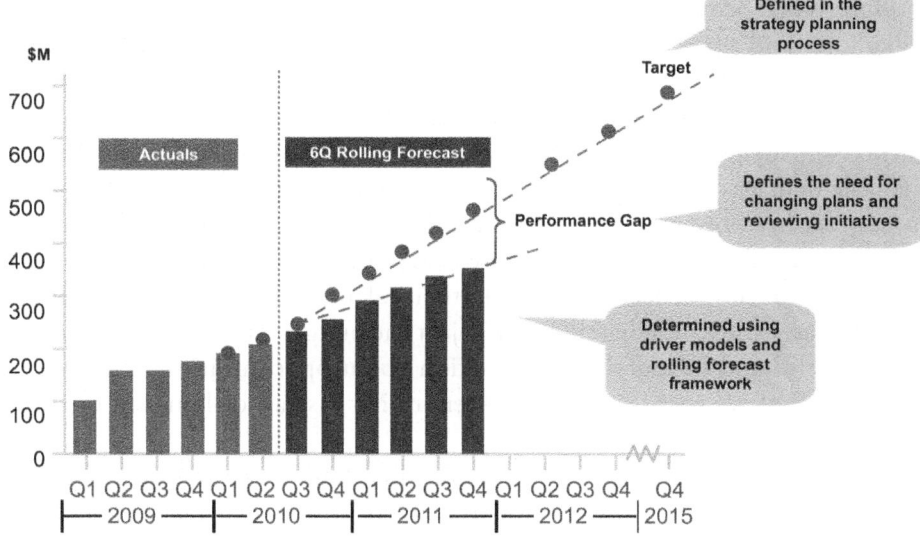

Fig. 7.4 Using rolling forecasts alongside Balanced Scorecard targets and initiatives. (Source: Palladium)

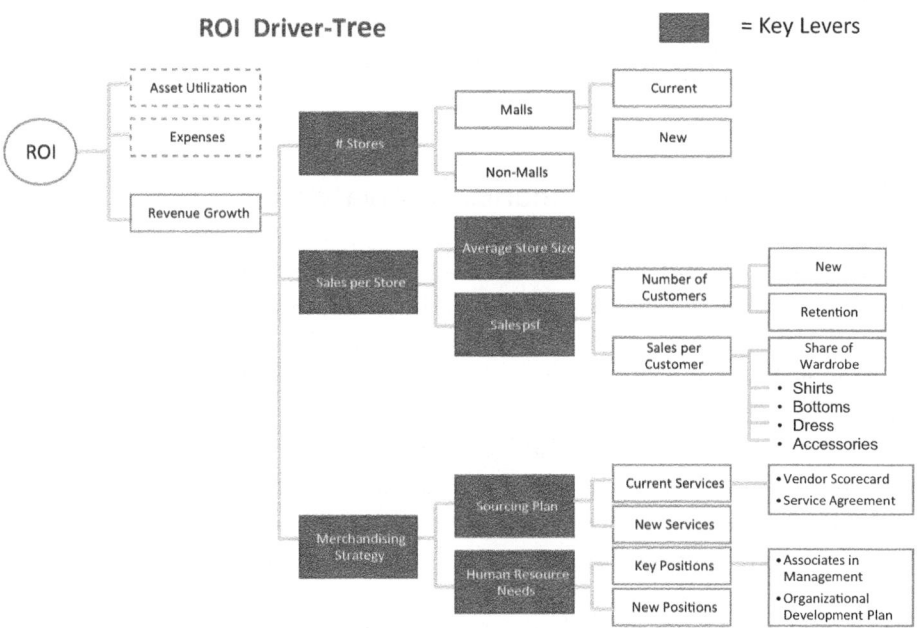

Fig. 7.5 Driver models provide the analytical framework to focus on key leverage points and to link operational KPIs and action plans to strategic priorities

Don Ryder, Senior Managing Director, Strategy Execution Consulting for Palladium (and a previous colleague of the authors of this book) provides this example. "Imagine asking your sales manager what she expects next quarter's Travel and Expenses to be for the sales team. She's going to make an educated guess. Now imagine that instead, you know that the trips the sales representatives she manages make are a primary driver of cost for the sales department, so you ask her how many trips her reps will make (the volume), which is a number she already uses to manage her department. She can give you these numbers with confidence, and you can then calculate the average cost of each trip (the rate) and immediately set or adjust the budget. What's more, you can have a useful and productive conversation about opportunities for cost savings – what if we spent a $100 less per trip? – instead of handing down a budget cut."

Ryder also makes the valid observation that many organizations treat the budgeting process as a substitute for strategic planning, which simply doesn't work. "There's a temptation to wrap everything together – financial planning, strategic planning, etc. – into a single package, but it ends up far too complicated and convoluted to be useful," he explains. "Typically, the short-term detailed budget takes so much attention that little, if any, strategic planning gets done."

He concludes, "We obviously need financial control mechanism to deliver on our financial commitments, but beyond that role the budget is the wrong tool to use."

Budgeting: Not a Performance Motivator

The conventional budgeting process is simply not fit-for-purpose for the digital age. As well as the shortcomings already listed, it also sends confused messages to employees. Through one channel, employees are informed that they are empowered, trusted, and so on, and through another channel, micromanaged through detailed budgets (not much empowerment and trust here). Of course, in any conflict, the budget always wins.

In many ways, it comes back to a simple question. "What is the most effective way to motivate better performance?" The answer is certainly not the annual budget. It never has been and is fast becoming a very dysfunctional approach to planning, management, and driving great performance in the digital age. It is now time to dance to a different tune.

> **Advice Snippet**
>
> The Beyond Budgeting Institute has formulated the following 12 principles (to both leadership and the management process dimension) to use when developing a more adaptive approach to financial management.
>
> **Leadership Principles**
> 1. Purpose – Engage and inspire people around bold and noble causes; not around short-term financial targets
> 2. Values – Govern through shared values and sound judgement; not through detailed rules and regulations
> 3. Transparency – Make information open for self-regulation, innovation, learning, and control; don't restrict it
> 4. Organization – Cultivate a strong sense of belonging and organize around accountable teams; avoid hierarchical control and bureaucracy
> 5. Autonomy – Trust people with freedom to act; don't punish everyone if someone should abuse it
> 6. Customers – Connect everyone's work with customer needs; avoid conflicts of interest
>
> **Management Processes**
> 1. Rhythm – Organize management processes dynamically around business rhythms and events; not around the calendar year only
> 2. Targets – Set directional, ambitious, and relative goals; avoid fixed and cascaded targets
> 3. Plans and forecasts – Make planning and forecasting lean and unbiased processes; not rigid and political exercises
> 4. Resource allocation – Foster a cost-conscious mind-set and make resources available as needed; not through detailed annual budget allocations
> 5. Performance evaluation – Evaluate performance holistically and with peer feedback for learning and development; not based on measurement only and not for rewards only
> 6. Rewards – Reward shared success against competition, not against fixed performance contracts

Linking Operations to Strategy

A decade has passed since Doctors Kaplan and Norton released the book that introduced the concept of the Execution Premium Process (XPP™). With the goal of offering a complete strategy management solution, *The Execution Premium: Linking Strategy to Operations for Competitive Advantage* served as Kaplan and Norton's final instalment in the ground-breaking series of books that began in 1996 with *The Balanced Scorecard: translating strategy into action*.

However, for all its logic and strengths, the XPP has rarely (if ever) been deployed as a single strategy execution framework. Rather, even when proudly displayed by organizations as their strategy management guide, only bits of the XPP are typically used. Most sequence straight from stage 3: *Align the Organization* (developing cascaded maps and scorecards) to stage 5: *Monitor and Learn*.

Yet, what distinguishes the XPP as a framework for linking strategy and operations is what happens between stages 4 and 5—the Execution phase. In retrospect, it might have been sensible to label Execution as a stage. It is taught as a "seventh stage" within the Kaplan and Norton Balanced Scorecard certification boot camp.

To align operations with strategy, Kaplan and Norton posed the question, "What business process improvements are most critical for executing the strategy?"

Strategically Critical Processes

Note the words "most critical." From our field observations and research, this is where organizations typically make a significant mistake in understanding and therefore implementation.

This stage is not about aligning *all* operations with strategy but, more specifically, those operational processes that have a significant impact on the strategic processes, as identified in the internal process layer of the Strategy Map and Balanced Scorecard. The criticality of improving strategic processes is precisely why Kaplan and Norton recommended that about 40% of objectives on the Strategy Map and accompanying Balanced Scorecard should be within the internal process layer. This is where work gets done and is the conduit to aligning operations.

The Dangers of Focusing on Continuous Process Improvement

Very often, organizations embark on huge process improvements efforts just for the for sake of improving, without a clear view of the strategic rationale for doing so. Process improvements are of course legitimate and desirable, but considering the limited resources available in any organization, it is advisable to concentrate efforts on those aspects with higher impact on the building of strategic capabilities.

This is not to say organizations ignore "non-strategic" processes. Many processes are vital, but not strategic—for instance, payroll payments—and they must work at an acceptable quality level. The strategic processes are those that produce greater benefits for the organization. Therefore, organizations should excel at those processes and function at an acceptable level for everything else.

The blind focus on being "good at everything" is partly the result of the legacy of Total Quality Management (TQM), which swept onto Western lands in the 1980s following the success of Japanese companies with TQM principles. Many organizations have become obsessed with continuous process improvement.

To a large extent, there's nothing wrong with this. The tools and techniques introduced by the quality gurus Doctors Deming, Juran, and others have made a massive and positive impact on improving organizational performance. Moreover, Deming's thinking played a significant role in moving away from F.W. Taylors, *Principles of Scientific Management*. For instance, Deming's principles of "driving out fear" and "restoring pride in work" were in sharp contrast to Taylor's approach.

However, there are downsides in overly focusing on continuous improvement.

A major criticism of continuous improvement efforts is that they are typically implemented bottom-up. As a result, they only look at tiny areas of their business and are not linked to a bigger strategic picture.

This means that teams typically pick the lowest hanging fruits, but often miss the big opportunities. Indeed, there are examples of organizations considered "masters" of continuous process improvement experiencing catastrophic failures. Motorola being a case in point (see below)—and its celebrated usage of Six Sigma. First, a short explanation of Six Sigma.

Six Sigma Explained

Developed by Motorola in the 1980s, Six Sigma is one of the most popular total quality type tools. At the core of a wider process improvement methodology, it has a goal of reducing error rates to 3.4 per million opportunities.

To illustrate what this means, consider the goalkeeper of a soccer team who plays 50 games in a season and who in each game faces 50 shots from the opposing team. A defect occurs when the opposing team scores a goal.

Therefore, a goalkeeper that performs at a Six Sigma standard would concede one goal once every 147 years! Just imagine such a level of performance.

As the soccer analogy illustrates, as a metric, it represents an extremely low defect rate. Moving from the soccer field to an organizational setting, for a business process, the "Sigma capability" is a metric that indicates how well the process is performing. Sigma capability measures the capability of the process to perform defect-free work. A defect is anything that results in customer dissatisfaction. As defects go down, the Sigma level goes up.

There are myriad examples of organizations that succeeded on a continual basis while using Six Sigma as its core approach to process improvement (such as GE, for instance). However, it is telling to note that some of the organizations that became poster-boys for Six Sigma and secured mouth-watering cost savings from their efforts, have simultaneously been very poor performers on the stock market and have been recognized for their strategic failures: Motorola being perhaps the most notable example.

Case Illustration: Motorola

Motorola reduced manufacturing costs by $1.4 billion from 1987 to 1994 and reportedly saved $15 billion over a period of about a decade. Impressive stuff! However, viewed from another perspective, it tells a very different performance story.

Motorola continually underperformed financially throughout the first decade of this century. Having lost $4.3 billion from 2007 to 2009, the company was divided into two independent public companies—Motorola Mobility and Motorola Solutions—on January 4, 2011. In August 2011, Google acquired Motorola Mobility for $12.5 billion.

The argument has been that Motorola, and other organizations, have been exclusively focused on using Six Sigma to identify cost saving opportunities rather than as a tool to continuously improve performance against a strategic goal. This is perhaps overly simplistic.

More broadly, many argue that Motorola simply made big strategic mistakes. Among the failures pointed to, we find, "Motorola missed the movement to 3G," "Motorola should have moved into content," "Motorola stopped innovating"—see below—"Motorola never had the sense of urgency and failed to grow." As one commentator notes, "Everyone else moves at warp speed; Motorola jogged at its own pace, more like a monopolist than a paranoid competitor." [3].

To put this into a sobering perspective, here is the company that invented the cell phone (it showcased the first hand-held portable telephone in 1973)—in the fastest growing market in all of technology—being clobbered in that market. Akin, perhaps to the inventor of digital photography, Kodak, being destroyed in that market (see Chap. 2: *From Industrial- to Digital-Age-based strategies*).

The Motorola story supports a further criticism of approaches such as Six Sigma, in that such efforts are "narrowly designed to fix an existing process" and therefore do not help in coming up with new products or launching/dealing with disruptive technologies.

Thwarting Innovation

Six Sigma has come in for some specific criticism as we move into the digital age, in which innovation is critically important. While concurring that Six Sigma has its value, A 2014 *Forbes* article, *Is Six Sigma Killing Your Company's Future?* cautioned that, "if you're not careful, innovation and growth may be swept away in the process." Seemingly, true of the poster boy, Motorola.

The article argues that because of Six Sigma or similar approaches, many organizations are today operating at very high levels of efficiency. However, as leaders now shift their focus to the acceleration of growth, they are discovering that the very culture of little to no variance that allowed them to achieve their efficiency goals is suffocating their growth potential. As the article's author comments, "Variance – dare I even say error – is essential for innovation and growth."

Supporting a message from Chap. 1 of this book (about evolution), the article goes on to say that variance is the very characteristic that dictates the rate at which evolution in nature occurs: "Consider the countless number of times that your cells divide to make you who you are. Now consider that each time a cell replicates, it must copy and transmit the exact same sequence of 3 billion nucleotides to its daughter cells. Inevitably, errors occur – and much more often than you might expect. Yet it is these errors in DNA replication that have allowed single-cell organisms to evolve into the unimaginably complex beings that we are today."

The article explains that a culture built on the foundation of eliminating variance can have dramatically negative effects on growth and innovation. "Experimentation, and the possibility of negative outcomes, becomes taboo. Managers are promoted based on having the right answers, not asking the right questions. Innovators are forced to create 'fantasy plans'—

with unrealistic and inflexible assumptions—to win approval for their projects. CFOs are paid to kill projects."

The article rightly argues that, "Those organizations who have successfully shielded themselves from variation and experimentation – those who have attained literal mastery of Six Sigma, ultimately face the greater risk of [evolutionary stagnation] …Variation is not an obstacle to steadfastly avoid; it is a key to unlocking your breakthrough future. Success requires embracing the alternative paths that markets randomly present to us in order to find your organization's own unique place to grow and thrive," [4]. In other words, being agile and adaptive.

Making the Strategy/Operations Link

Although being overly focused on continuous improvement to the detriment of strategy is a major issue, (according to Hackett's assessment criteria) this is oftentimes compounded by some of the approaches taken to cascade Balanced Scorecards.

We have observed many organizations that labour under the false belief that aligning operations is about creating Balanced Scorecards at devolved levels (and sadly, many consultants preach this), where operational indicators can predominate. It is not.

Devolved scorecards are the work of aligning the organization, an earlier stage. Moreover, operational indicators should appear on operational dashboards—not Balanced Scorecards, which house strategic measures. Confusing operational dashboards with departmental scorecards leads to a diluting of focus on the strategically critical processes, with improvement interventions becoming tactical and not strategic. Strategically important improvement opportunities are overlooked or not prioritized.

To select strategically important improvement opportunities, organizations must first distil the internal process objectives into sub-processes, thus making it easier to identify specific, strategically aligned interventions.

Then, and only then, can applying Six Sigma, total quality, re-engineering, and/or other efforts to improving performance deliver significant, and oftentimes quick, benefits. Such interventions are shorter term and less resource intensive than major initiatives, which often span many processes and even across perspectives, and from this help deliver the required customer and financial outcomes that appear in the top half of a Strategy Map.

Driver-Based Models

This is where we use driver-based models (which asks the question, "What must we excel at to deliver to this objective?") to identify the key sub-processes.

As explained in Chap. 4, *Strategy Mapping in Disruptive Times*, we recommend the use of a well-written objective statement as a link between the strategic objective and the driver-based model and, from that, the KPIs. This enables a more focused steer for the identification of the sub-processes.

Case Illustration: Hospital Complex

In Chaps. 4 and 5, we provided the example of how a hospital complex used an objective statement to identify three value drivers for delivering to the objective "Assure service excellence & optimize the customer experience."—patient access, care coordination, and patient discharge.

A further driver-based exercise distilled the sub-processes to take the objective further down the operational chain. Figure 7.6 shows the operational processes, operational drivers, and measures for the sub-process "patient access." Performance improvements were identified and monitored through measures on an operational dashboard (not a departmental scorecard). Figure 7.7 shows the patient access dashboard.

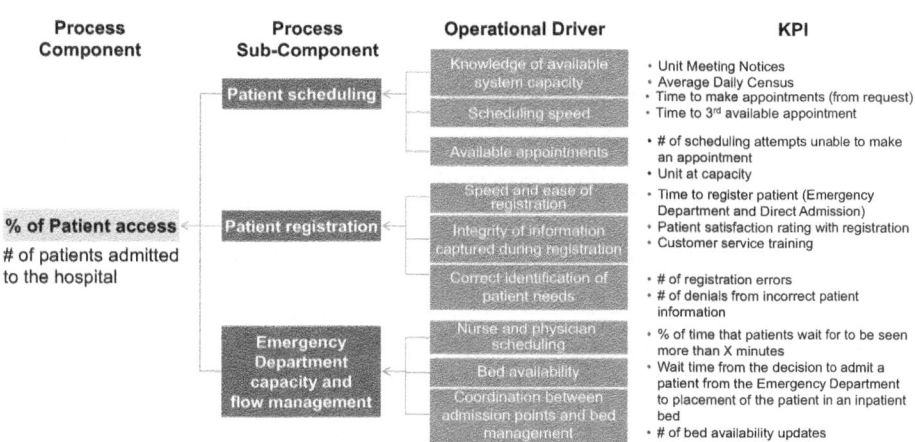

Fig. 7.6 Decomposing an internal process objective into operational drivers and KPIs

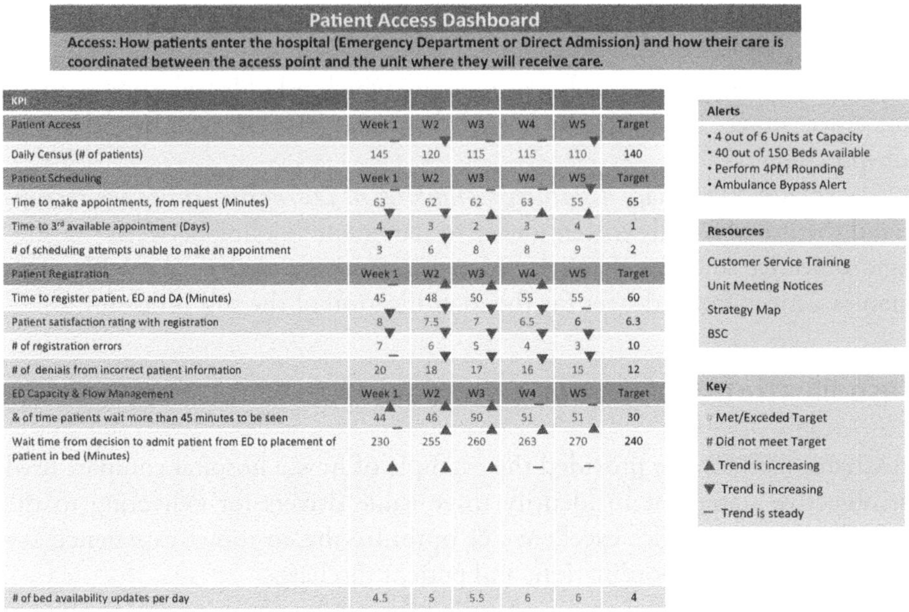

Patient Access Dashboard
Access: How patients enter the hospital (Emergency Department or Direct Admission) and how their care is coordinated between the access point and the unit where they will receive care.

KPI	Week 1	W2	W3	W4	W5	Target
Patient Access						
Daily Census (# of patients)	145	120	115	115	110	140
Patient Scheduling	Week 1	W2	W3	W4	W5	Target
Time to make appointments, from request (Minutes)	63	62	62	63	55	65
Time to 3rd available appointment (Days)	4	3	2	3	4	1
# of scheduling attempts unable to make an appointment	3	6	8	8	9	2
Patient Registration	Week 1	W2	W3	W4	W5	Target
Time to register patient, ED and DA (Minutes)	45	48	50	55	55	60
Patient satisfaction rating with registration	8	7.5	7	6.5	6	6.3
# of registration errors	7	6	5	4	3	10
# of denials from incorrect patient information	20	18	17	16	15	12
ED Capacity & Flow Management	Week 1	W2	W3	W4	W5	Target
% of time patients wait more than 45 minutes to be seen	44	46	50	51	51	30
Wait time from decision to admit patient from ED to placement of patient in bed (Minutes)	230	255	260	263	270	240
# of bed availability updates per day	4.5	5	5.5	6	6	4

Alerts
- 4 out of 6 Units at Capacity
- 40 out of 150 Beds Available
- Perform 4PM Rounding
- Ambulance Bypass Alert

Resources
- Customer Service Training
- Unit Meeting Notices
- Strategy Map
- BSC

Key
- Met/Exceeded Target
- # Did not meet Target
- ▲ Trend is increasing
- ▼ Trend is increasing
- — Trend is steady

Fig. 7.7 Patient access operational dashboard

As a result, improvements at the operational level could be tracked to the progress of the strategic measures on the internal process objective of the Strategy Map, and therefore through to the corresponding customer and financial objectives. Strategy and operations are linked, but managed separately.

Advice Snippet

Do not confuse the strategic scorecard with operational dashboards.

- *Strategic scorecards* are not about detailed day-to-day activities, but future oriented goals and creating a map to get to somewhere that the organization hasn't been before.
- *Operational dashboards* are about checking that things are going right and, if not, then doing something about it. It is near real-time checking and operational monitoring.

Operational dashboards can be effective tools for driving performance improvement. The linkage to strategy is when the dashboard elements are:

- Derived through explicit decomposition of strategic objectives within the internal process perspective, in order to identify the operational drivers
- Designed to facilitate decision-making tools for a specific set of users
- Actionable – the activities necessary to improve performance are clear and unambiguous.

Parting Words

"Successful strategy execution has two basic rules: understand the management cycle that links strategy and operations, and know what tools to apply at each stage of the cycle," say Doctors Kaplan and Norton.

A key element of linkage is ensuring the use of the appropriate tools for execution. In this chapter, we considered the application of tools for process improvements, while in the next chapter we explain how to manage and implement breakthrough strategic initiatives, which research has shown to be a major issue for most organizations and a key reason for supposed Balanced Scorecard "failures."

Panel 1: Practitioner Advice on Moving on from Traditional Budgeting

Within an edition of Palladium's *Strategically Speaking (edited by one of the authors of this book)*, Andreas de Vries, Oil & Gas strategy management specialist, provides this useful advice on evolving financial management for the digital age.

"From a strategic perspective, the value in the traditional budgeting process is essentially control. Based on an expectation regarding the future, the organization identifies the actions it believes will enable the achievement of its objectives. These actions are then priced to develop a budget, after which management regularly reviews actual expenditures in light of this budget to ensure only the pre-defined actions are executed – there isn't a budget for anything else, after all."

A Heavy Price Tag
"This value comes at a substantial cost. Most organizations use a considerable amount of resources for a substantial portion of each year (usually, anywhere between three to nine months) for budgeting. Greater still is the cost that comes from the tendency of budgets to lead organizations towards doing what shouldn't be done, while preventing them from doing what should be done. The traditional budgeting process locks organizations into executing a pre-defined list of actions, as it doesn't have a mechanism to adjust the action plan if, and when, circumstances change in unforeseen ways. What we tend to see, of course, is a significant re-budgeting effort half way through the year."

A New Budgeting Process
"In this day and age, however, unforeseen events are the rule rather than the exception. This rapidly changing business environment therefore requires a new budgeting process, one that maintains control over the traditional budgeting process to facilitate considered resource allocation decisions, but at the same time provides the flexibility needed to navigate through a sea of ever-changing threats and opportunities. I recommend three things in particular."

Be Less Detailed
"First, since it should be expected that the business environment will not develop exactly as foreseen during the planning and budgeting exercise, organizations should stop detailing their budgets to the nth degree. In an unpredictable business environment, this is a waste of time, effort, and resources."

Budgeting Is Not Management
"Second, management should stop equating budgeting with management. The days of "as long as we follow the budget, we'll be successful" are over. The objective of budgeting should be to force an organization to think about the future, to develop preliminary action plans that would make it successful in that future, and to make a preliminary decision on resource allocation that would maximize returns in that future—not a final decision."

Update Regularly
"Third, updates of the original budget should be developed regularly, based on the actions deemed necessary to achieve the objectives at that particular point in time. The expenditure reviews should then focus on understanding the reason for any differences between the original and updated budget: *What changed in the business environment so that we now believe we have to do different actions?* This shifts the focus of expenditure analysis from *Have you done what you said you would do?* to *Are you doing what needs to be done?* Only thereafter should the final decision on resource allocation be made, as close as possible to the actual event" [5].

Self-Assessment Checklist

The following self-assessment assists the reader in identifying strengths and opportunities for improvement against the key performance dimension that we consider critical for succeeding with strategy management in the digital age.

For each question, any degree of agreement to the statement closer to one represents a significant opportunity for improvement.

Please tick the number that is the closest to the statement with which you agree		
	7 6 5 4 3 2 1	
In my organization, financial planning is very well aligned to strategic planning		In my organization, financial planning is very poorly aligned to strategic planning
In my organization, the budget is subservient to strategy		In my organization, strategy is subservient to the budget
Generally, managers are very satisfied with the budgeting process		Generally, managers are very dissatisfied with the budgeting process
In my organization, financial resources are generally deployed when needed		In my organization, financial resources are generally locked in for the duration of the budget

We use rolling forecasts that look beyond the end of the financial year	We forecast only to the end of the financial year
Incentive compensation is generally decoupled from the annual budget	Incentive compensation is generally hardwired to the annual budget
In my organization, we have a very good understanding of which business process improvements are most critical for executing the strategy	In my organization, we have a very poor understanding of which business process improvements are most critical for executing the strategy
We have a very good process for reviewing how long-term objective and short-term performance interact	We have a very poor process for reviewing how long-term objective and short-term performance interact
In my organization, operational performance is monitored through operational dashboards	In my organization, operational performance is monitored through the same scorecards/dashboards as strategy
We are very good at using driver-based models, or similar, to align operational processes with strategy	We do not use driver-based models, or similar, to align operational processes with strategy

References

1. Lori Clabro, *On Balance*, CFO Magazine, February 2001.
2. Jack Welch, Suzy Welch, *Winning*, Harper Collins, 2005.
3. Howard Anderson, *10 Reasons why Motorola Failed*, Network World, 2008.
4. Rick SMith, *Is Six Sigma Killing Your Company's Future?* Forbes, June 2014.
5. Andrease de Vries, *Is the Budgeting Process Still Fit-For-Purpose*, Strategically Speaking, Palladium, August 2015.

8

Developing Strategy-Aligned Project Management Capabilities

Introduction

Repeated research has found that successfully implementing strategic initiatives is notoriously difficult to achieve and is a major reason why even the most carefully developed Strategy Maps and Balanced Scorecards can fail to deliver expected results (Fig. 8.1).

One reason for this is when reviewing how organizations build frameworks—such as the Balanced Scorecard System—to implement and manage strategy, we note that the general approach is to shape the Strategy Map and scorecard (oftentimes taking a long time to do so), and then moving straight to the quarterly reporting. The "bit," in the middle, the actual execution of the chosen initiatives (along with incremental process improvements, [see previous chapter]), does not receive the same rigorous attention. It's almost as if once the scorecard system has been built, the rest will automatically take care of itself. It will not.

Not an Execution Framework

Here we find a common misconception of the Balanced Scorecard system. It is *not* a strategy execution framework, as it does not implement anything. It is a framework for translating strategy (the chosen objectives through to initiatives). Kaplan and Norton's Execution Premium Process (Fig. 8.2) has execution as a sort of "seventh" step that is in between stage 4, "align operations," and stage 5, "monitor and learn." Not being an official stage (i.e. with an

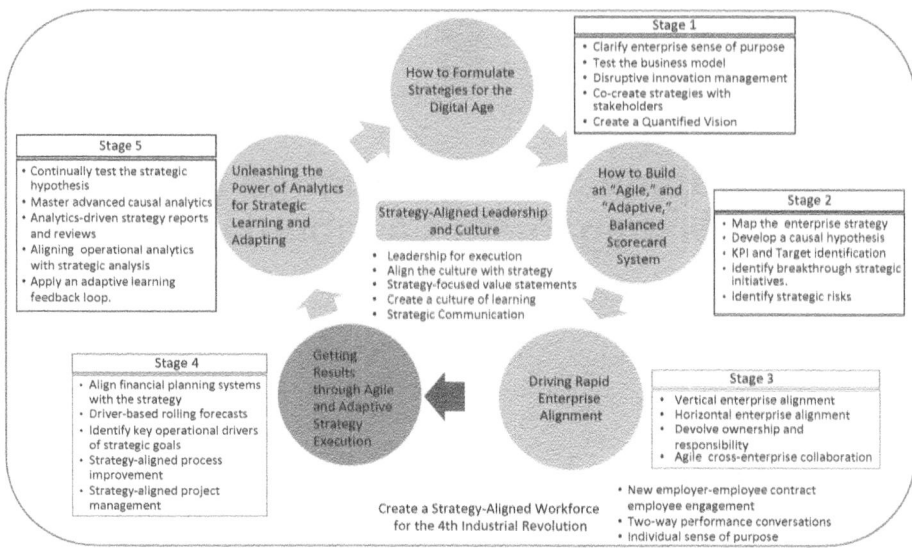

Fig. 8.1 Stage 4: getting results through agile and adaptive strategy execution

Fig. 8.2 Execution: the "seventh" stage of the execution premium process

assigned number) has led to it being generally overlooked, or more truthfully, disregarded. Build the system and then report. This is why we have incorporated it into Stage 4, *Getting Results through Strategy Execution*.

Professional Expertise

Now, there might be one key reason for this oversight regarding "stage 7." Initiative management takes us into the well-established world of project management, which has robust and accepted rules and processes (such as the Project Management Institute's [PMI] Body of Knowledge), benefiting from highly regarded and sought-after qualifications, such as PMI's Project Management Professional (PMP), Program Management Professional (PgMP), and Portfolio Management Professional (PfMP) standards.

Agile Project Management

In an explanation of where being agile and adaptive are most relevant in the strategy management cycle, we must give some time to agile project management. Growing out of *the Agile Manifesto* in 2001 [1], which sought to deal with the then rising frustration with failed software development projects (see Chap. 2: *From Industrial- to Digital-Age*-based strategies), agile project management essentially focuses on continuous improvement, scope flexibility, team input, and delivering quality products. An agile approach differs from a conventional approach because a more rapid decision-making process replaces the traditional decision-making hierarchy (project manager, sponsor, steering review, etc.)

PMI states, "Agile approaches to project management aim for early, measurable ROI [Return on Investment] through defined, iterative delivery of product increments. They feature continuous involvement of the customer throughout product development cycle." Table 8.1 describes the key differences between a traditional and agile approach to project management.

SCRUM

A specialized agile project management method is SCRUM (taken from the SCRUM in Rugby Union). On a project, SCRUM enables faster decision making and involves two key roles: Product Owner and the SCRUM Master, supported by an expert team.

Table 8.1 The difference between traditional and agile approaches to project management

Traditional approach	Agile approach
Customers get involved early in the process but tend to be kept at arm's length once the project has started	Customer involvement is the key here as more closely involved at the time that work is performed
Escalation to managers when problems arise	When problems occur, the team works together to resolve them internally
Heavy upfront analysis and design	Daily stand-up meetings are held to discuss the work done yesterday, the plan for today, and impediments if any
More serious about processes than the product	Agile methods focus less on formal and directive processes
Product is planned extensively and then executed and tested	Work is delivered to the client in small and frequent releases to get rapid feedback loops
Traditional model favours anticipation	Agile model favours adaptation
Heavy documentation before executing the work packages	Agile projects are highly democratic and implement a series of short and repeatable practices
Slow and structured development process, usually focused on a one-time release	Agile is more focused on "why," with retrospection at the end of each iteration or release
Belongs only to the project manager	Shared ownership

The Product Owner (customer) is responsible for the outcome of the project and is involved throughout the process, aiding more rapid decision making and ensuring that the project is delivering to requirements.

The SCRUM Master helps team members work together in the most effective manner possible, breaking down and assigning tasks to team members to implement so as to deliver the product solutions. The SCRUM Master coaches the team, removes impediments to progress, facilitates meetings and discussions, and performs typical project management duties such as tracking progress and risks.

Central to the SCRUM method are short stand-up meetings (called Sprints) which consider achievements and lessons learned since the last meeting and focus on what needs to be done before the next. These Sprint meetings are integral to how the SCRUM method delivers specified solutions to meet the project milestones.

Moreover, the belief is that successfully meeting regular goals energizes the team and maintains positive momentum. US-based Certified Scrum Professional James Bass comments, "The value of SCRUM is not measured in the amount produced, but on the success of the team to meet commitments, and continuously improve. In this way, SCRUM provides value both to the team and to the organization through ever improving velocity, team empowerment, and success through controlled failure."

We will now consider what agile project management means for strategy execution. Saliha Ismail, Head of Strategic Projects within the PMO of the Ministry of Works, Municipality Affairs & Urban Planning, Bahrain (and a Certified Kaplan/Norton Graduate), explains that to best use agile in strategy execution, it is important to think about who the Product Owner and the SCRUM Master will be. "The Product Owner should be a senior executive of the organization, as they have the authority to make decisions that might cut through various functions, and the SCRUM Master should perhaps sit within the strategy department."

She continues that the SCRUM Master monitors whether performance has improved because of the initiative and so is in charge of moving the dials against the strategic KPIs. "Based on this data the Product Owner can then decide whether the initiative is delivering to outcome expectations as is, requires some modification or be abandoned/scaled down or whether the KPI itself is still valid."

Ismail's argument that placing the SCRUM Master within the strategy department or Office of Strategy Management (OSM) is useful for one key reason: it helps mitigate the regular PMO complaint that their territory is being invaded by unqualified people from the OSM, who claim that they are owners of strategic initiatives. We have witnessed many cases where the ensuing friction has led to the PMO and OSM refusing to talk to each other!

As well as leading to a sub-optimized initiative management process, it invariably means that beleaguered managers in the field have to report performance of the same project/initiative through two different systems—a strategy management system and project management information system (which is common even when they do collaborate). A further dysfunctional outcome is that the PMO and the OSM interpret, and so report, the results in oftentimes very different ways, to the continued ire of the senior team.

The Case for Merger

A simple solution to this issue is to merge the two offices, which would lead to the SCRUM Master being in the "strategy" department. As one Chief Strategy Officer said to one of the authors of this book, "When I took over it was obvious they weren't working together so I merged them – it was a no brainer!"

"The PMO's expertise is around ensuring that projects/programs are managed efficiently and to established procedures – be they strategic initiatives or large tactical or operational projects. An OSM is primarily required to manage the strategy process and facilitate strategic alignment as well as the review and update of the strategy," continues Ismail. "Clear delineation of roles is required, even when capabilities are merged." That said, PMI has, in recent years, made a strong play for PMOs to be repositioned/upskilled as the natural custodians of managing strategic initiatives, as explained in Panel 2.

> **Advice Snippet**
>
> If the organization has a PMO and an OSM, consider their merging. Both offices are working toward the same goal—delivering the strategy. But, write charters for both that clearly describe who does what in the process.
>
> In many organizations, it makes sense to combine the PMO, OSM, and quality management departments, as it enables a stronger link between strategy and all the action required for its delivery: execution becomes a joined-up, single process.

The Transformation Office

A particularly useful idea that has recently emerged is that of a Transformation Office (the capabilities for which can also be integrated into the merged PMO/OSM). Introduced by the Gulf-based consultancy ShiftIn, the need for such a capability was explained in a 2016 white paper, *"Creating the Transformation Office,"* co-written by one of the authors of this book [2].

Transformation: A Definition

According to the paper, transformation "encompasses the adaptation of an organization's value proposition and business & operational model to the fundamental changes brought by global trends, digital technologies and increasingly demanding stakeholders' needs and expectations."

Transformation was categorized into three core verticals and three transversal transformations:

Core Vertical Transformations

- **Turnaround / Financial**: "The transformation that is driven mainly by a seriously underperforming business or organization: as much as anything a survival imperative."

- **Mandate / Core:** "The transformation that is driven by changes in policy, mandates or of the vision of the senior top leadership (e.g., board, owner, heads of government)."
- **Business Model:** The transformation that is driven by fundamental changes on the value proposition and/or the way that products/services are delivered, the markets that are served, and the cost/revenues (or value) structures.

Transversal Transformations

- **Digital:** "The transformation driven by digitalizing the organization's DNA as well as key aspects (partially or fully) of its value proposition."
- **Operational:** "The transformation driven by changes at the core of the operational processes in pursuit of efficiency and optimization, usually deep rooted in supply change transformation."
- **IT & Technology:** "The transformation of the current technology and IT backbone of the organization, driven by an increase on demands from the business (or social) environment."
- **Support functions:** "The transformation of legacy structures in terms of support (such as legal, HR, finance) driven by an increase on demands from business (or social) environments as well as the continuous seek for efficiency, cost reduction and service level increase."

The paper's authors stress that in practice, there will likely be no clear boundaries between these transformations and, most likely, a core vertical transformation is pegged by a good share of transversal transformation. "For instance, the change in a business model, where an entire new value proposition will be introduced (e.g. moving from physical products to digital services), will be dependent on a set of internal transformations in terms of digital, operational and IT components."

Transformation and Strategy

Transformation initiatives always anchor to the longer-term strategy and (typically mid-term in nature) are the key strategic initiatives that deliver breakthrough performance improvement. This supports the importance placed by the authors of this book on mid-term plans (see Chap. 4: *Strategy Mapping in Disruptive Times*).

In the paper, the authors argue that, for many organizations, the requirement to transform or fail (even die) is so critical that there is a pressing need to re-think how organizations manage such mission-critical initiatives. "Transformation initiatives bring an additional layer of complexity that other projects do not and often have a deep disruptive impact on the organization's prevailing structural and cultural dynamic. Transformation efforts can (and most likely will) fail if is not properly managed as a unique set of initiatives…therefore there is a pressing need for a dedicated Transformation Office, responsible for driving such complex, disruptive change initiatives."

The paper's authors do not suggest that all organization necessarily require three standalone offices. Organizations' preferences regarding structure and culture will come into play. "What is more important is to recognize the distinct, yet complementary roles of the three offices. Each require its own remit and charter, whether merged or not."

By being fully cognizant of the specific requirements from all three offices then joined-up, impactful end-to-end enterprise performance management becomes achievable (from short-term operational planning, through mid-tern transformation planning, to long-term strategic planning). Confusion, turf-wars, and overlapping of efforts are eliminated, with the three offices laser-focused on what it must deliver and how this positively impacts the organization more broadly. Figure 8.3 shows the different responsibilities for the three offices.

The paper concludes by arguing that the demands of the "digital economy" are heavy and unprecedented (supporting a theme of this book—how organizations will be reconfigured going forward is still a work in progress—and will be for some time). "But what is already glaringly evident is that agility and the required focus on transformation will become hallmarks of the most successful organizations – and increasingly a perquisite for survival. Building the structures, governance and process models that make transformation a core organizational capability will become an organizational 'must-have.' A dedicated Transformation Office, that is distinct to a more conventional Project Management or Strategy Office, is…a critical step in building this must-have capability."

There are numerous variations in how strategy and projects are aligned. An interesting spin is the Turkey-based telecommunications provider, Turkcell Superonline, which has also integrated making into its OSM, as we explain in Panel 1.

Developing Strategy-Aligned Project Management Capabilities

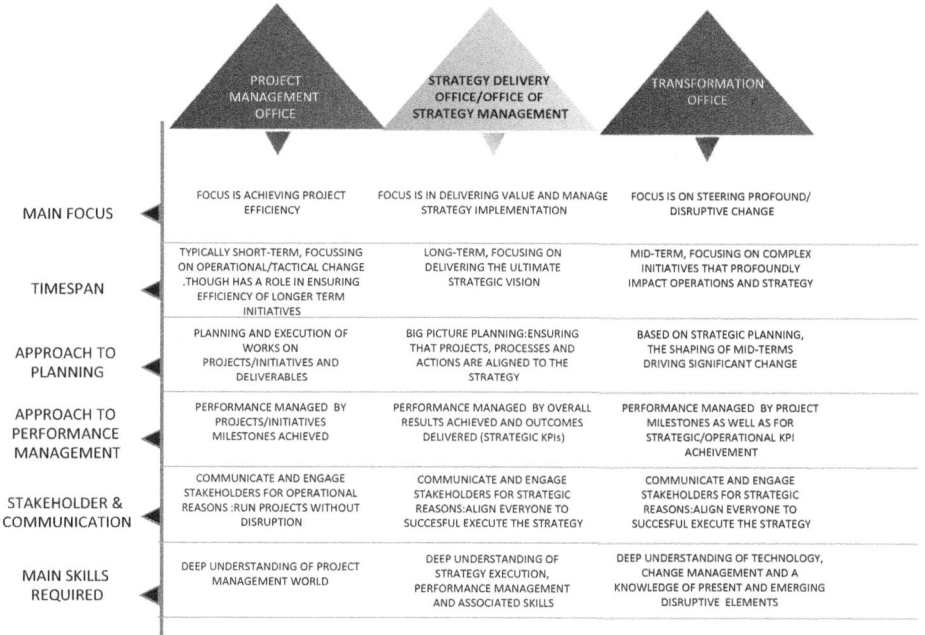

Fig. 8.3 The roles of the three offices within a transformation office. (Source: ShiftIn)

Measuring Impact

But, however configured, one area that organizations often struggle with that still needs to be redressed is how to measurer or assess the impact of initiatives. As we explained in Chap. 5: *How to Build an agile and adaptive Balanced Scorecard,* the purpose of a strategic initiative is to close the performance gap (from "as is" to "to be" against KPI targets). Collectively, these close the value gap (from the "as is" to the "to be" states against a quantified vision—see Chap. 2: *From Industrial- to Digital-Age-based strategies.*) Classic project management measures of time, cost, and scope are, of course, required, but do not in and of themselves measure value or impact, a point elaborated on by James Coffey, Principal of the US-based consultancy Beyond Scorecard, who highlights two aspects of initiative management.

"First there is the project management aspect of tracking – is it on time, on budget, what are risks, etc. In other words, the mechanics of project management, which is best left to a PMO or manager.

Second, there's the answer to the questions: "What is it intended to accomplish, and will it achieve its goals?" The leadership team answers the first part of the question when choosing initiatives to support strategic goals. The answer to the second element requires periodic revisiting to ensure it will actually do what is intended."

This, Coffey explains, is more complex than simply managing the project. It requires analysis of the risks, shifts in the landscape since the initiative started, impact of other initiatives, and so on. "It is an iterative discussion with the sponsor, the project manager and other managers to understand what is actually happening," he says, adding that "On time and on budget is not enough as this doesn't address whether the initiative will actually do anything other than be completed."

Two Types of Initiatives

For monitoring purposes, it is also worth keeping in mind that there are two types of strategic initiatives.

The first type is AFE (After Finalization Effects), for initiatives that have the realization period starting after project finalization.

The second type is DPE (During Progress Effects), for initiatives that have the realization period overlapping a part of the project timeline (but with a realization time lag).

An AFE initiative example might be installing a new CRM system. Its realization period is related to the time required for a significant number of customers acknowledging (or not) the benefits derived from the new CRM system's functions and capabilities. This cannot be done until the system is up and running.

An example DPE initiative is a retail outlet revamping based on a new concept. Its realization period overlaps the project timeline, as the customers entering the first newly revamped outlets are more (or less) satisfied with the new look and features, while the revamping project is on going in other locations. The initiative realization is the propagation (to outcome effects, sensed by the lag KPIs) of the initiative's immediate output effects (sensed by the lead KPIs). The realization time lag is the time required for turning the initiative output effects into the desired outcome effects. This time lag is usually never zero (Fig. 8.4).

A Portfolio Approach

A further common mistake in initiative management viewing initiatives as standalone: a portfolio approach is required. As Dr. David Norton comments, "Initiatives should be seen as a portfolio, as it is the collective impact that ultimately closes the value gap."

Fig. 8.4 Initiative realization time-lag. (Source: Synergys)

A 2013 report by Boston Consulting Group and the Performance Management Institute (see also Panel 2) explained the power of a portfolio approach. "[A] critical aspect of tracking progress is…proving a current portfolio approach for senior leaders."

The report's authors noted that an analysis of about 2000 initiative roadmaps, with $4 billion impact-bearing milestones found that 35% exceeded plan, 45% were within plan, and 20% fell short. "Overall, this portfolio of initiatives was successfully delivered, achieving more than 100% of the targeted value."

The report added, "A portfolio view…ensures that organizations adopt an enterprise-wide perspective. This generates the necessary forward-looking clarity on how the overall program is proceeding against planned impacts, spotlighting areas that require additional attention and enabling organizations to over-deliver against targets, despite inevitable and initially unforeseen implementation shortfalls in individual roadmaps."

Sub-Portfolios

If the overall set is a portfolio, then sub-portfolios can be created at the theme level. With the value gap defined, leaders then need to agree on how to close that gap. If, for instance, reaching the target requires significant improvement in new product development, then that theme would get the bulk of the initiative investment. Conversely, if, for example, there's a requirement to ease back on revenue building and to focus on efficiency for a while (as was the case with many financial institutions in the wake of the financial crisis), then investments can be switched to initiatives that deliver operational excellence. The map itself doesn't necessarily change, but the performance improvement efforts are geared to one theme in particular. Over the lifetime of the strategy, all themes are of equal importance, but investments are prioritized sequentially. One of the benefits of managing by themes (see also Chap. 4) is that it provides a powerful vehicle for investing along a causal chain.

Parting Words

The progress of strategic initiatives should be formally assessed as part of the strategy review calendar (typically quarterly, or perhaps, as in the Turkcell Superonline example in Panel 1, to a monthly schedule in fast-moving markets). We will consider the strategy review in the next chapter.

Panel 1: Turkcell Superonline Case Illustration

A wholly owned subsidiary of the Turkcell Group, Turkcell Superonline is the leading communications and technology company in Turkey. With market leadership in five of the nine countries in which it operates, Superonline had approximately 71.3 million subscribers as of December 31, 2013.

A Case for Change: First Wave – Building the Fiber Road (2008–2012)
A vibrant economy together with a sizeable and young market with very high Internet usage potential and low broadband penetration were the key factors that pushed the company to fundamentally shift its strategic focus.

The CEO and top executives led the first wave—transforming the then-named Tellcom from a wholesale telecom service provider to a telecom operator with its own next-generation fiber infrastructure, which started in 2008. In the ultra-fast broadband market that Tellcom was about to enter, the target segment was no longer just the wholesale business, but also households and organizations.

As part of the shift, in 2009, Tellcom merged with Superonline, one of Turkey's leading Internet providers. Turkcell Superonline initiated the "fiber age" in Turkey, offering Internet service at a speed of 1000 Mbps: the most advanced Internet access technology in the world, then provided in only seven other countries.

Establishing the OSM
As a further key element of driving transformational change, in 2008, the CEO introduced an Office of Strategy Management (OSM), reporting to him and with responsibility to facilitate and coordinate the strategy management process and taking control of driving strategy and change.

Seven focus areas were identified and strategic measures (superindicators) defined to assess progress. The top ten strategic initiatives were defined and aligned to the Strategy Map. Strategy Maps were developed for the three market segments: residential, corporate, and wholesale. Cascading also reached the functional and support units.

Creating an Office of Strategy Management and Strategic Marketing

At first, the Strategic Planning Department was empowered to lead the scorecard effort with the critical involvement of the CFO. However, this department was renamed the Office of Strategy Management and Strategic Marketing in 2011 and gained more responsibilities to execute, coordinate, and facilitate the whole strategy management process. The department started to report to the CMO as well, since the customer was a strategic focus in the next strategy wave instead of infrastructural expansion; thus, the department created synergies with the marketing team to support fast and accurate decisions.

Monitoring Performance

With responsibility for strategic initiatives within the OSM (and as set out in its OSM Charter), what is particularly innovative about the Turkcell approach is that strong customer analytics is seen as a strategic differentiator and is used extensively to monitor performance to, and the impact of, the small number of strategic initiatives that drive transformational change.

The organizations correctly realized that in the fast-moving sector in which it operates, quickly capturing shifting market/customer trends is a success prerequisite, and so, strategic initiatives must be constantly tracked for relevance and tweaked accordingly.

Turkcell Superonline monitors its strategic achievements at an especially intense level. The executive team holds weekly operational reviews for the day-to-day business issues and meets monthly and quarterly to review strategic performance (there's a twice-yearly formal strategy refresh).

In the monthly meetings, the accent is on the superindicators and the ten strategic initiatives that drive most of the strategy's success. Every quarter, competitors' behaviour is analysed and strategic actions and priorities (and therefore initiatives) are evaluated according to the environmental changes. Finally, there is an annual strategy review to adjust the strategy execution roadmap.

Second Wave: From "Good to Great" in Customer and Operational Excellence (2013–2018)

In 2012, the CEO announced a second wave of transformation, aimed at becoming "the indispensable technology partner by easing the life of our customers with innovative solutions in the digital world." The new vision involved switching from the infrastructure orientation to a customer-oriented approach. Emphasis on customer and operational excellence themes shaped the new strategy. Success is measured in terms of take-up rate, customer satisfaction and retention, profit, and cash flow. The company set the key milestones to accomplish the return of investment in accordance with customer satisfaction. Turkcell Superonline revised its strategy maps, KPIs, and strategic initiatives to reflect the second wave directions.

Key elements of wave two included the development of a more systematic initiative management system to better link budgeting, resource allocation, and decision making to the strategy and enhance the follow up on the initiatives, as well as the introduction of the scenario-based strategic planning for a more accurate and agile adaptation of Turkcell Superonline's strategy in an increasingly competitive landscape.

Measures of Success

As for success, consider the following. All figures compare 2008 with 2013:

- Total revenue increased fourfold, from 158 to 925 million Turkish Liras in four years.
- EBITDA turned from negative 42 million to positive 238 million Turkish Liras.
- Network coverage jumped from 64,000 homepass (fiber infrastructure coverage) users to more than 1.7 million in just four years.
- Market share in terms of fixed Internet subscribers (including ADSL, fiber, and cable Internet) grew from 3% in 2010 to 11.3% in 2013.
- Almost 60% decrease in call centre usage from 2010 to 2012, showing the improvement of the operational activities.
- Employee satisfaction rate was 16% above the average value in Turkey.

Abridged from Palladium's 2013 Hall of Fame Report and other materials. [3]

Panel 2: The PMOs Claim for Managing Strategic Initiatives

In 2013, the Project Management institute (PMI) released a report, jointly authored with Boston Consulting group, called *Strategic Initiative Management: the PMO Imperative*. The report argued that "A program or project management office (PMO) can play a crucial role...in actively supporting the implementation of strategic programs." However, it underscored the fact that, "the role of the PMO within the firm must, in turn, become more strategic, and it must develop its capabilities accordingly."

The report summarized four key skills required by PMOs, which resulted from several years of research among PMO and other leaders.

Focus on Critical Initiatives

Here, the message is to provide senior leaders with true operational insight through meaningful milestones and objectives for critical strategic initiatives. "Such focus provides clarity and course-correction among emerging issues and the fullest possible realization of impact."

The report stated that generally PMOs do already encourage the setting of milestones, measures, and objectives for initiatives, "but in many cases the traditional approaches simply aren't effective for large, complex and fast-moving initiatives," it was noted. "They generate either too little or too much information, obscuring what is really happening instead of providing PMOs and senior executives with clear information that allows for timely course correction."

> **Institute Smart and Simple Processes**
> This component spotlighted the importance of establishing program-level routines that track initiatives and their goals, communicate progress, and help to identify issues early without adding undue burdens or usurping the businesses and functions executing the work.
> "Roadmaps must identify the set of critical milestones that provide leadership with a clear basis for operational insight into what the initiative is about, what the critical known risks and interdependencies are, and how the initiative is progressing, using forward-looking indicators."
>
> **Foster Talent and Capabilities**
> Developing and nurturing the right technical, strategic, and business management, as well as leadership, skills, and capabilities within the organization was the next component.
> The report stated, "It is no longer enough to focus talent development only onto technical project-management skills. Organizations also need to develop a portfolio of talented PMO staff with strategic, business and leadership skills."
>
> **Encourage a Culture of Change**
> This component stressed the criticality of actively building organization-wide support for, and commitment to, strategic initiative implementation and changing management as real competitive differentiators.
> "The implementation of initiatives is hard, and much can go wrong during the journey...PMOs need to foster the transparency about problems as they come up, providing senior leaders with the means to make course corrections in time to ensure that the overall initiative hits its target impact," the authors state. "That in turn requires an environment in which people, especially business managers executing initiatives, are willing to raise their hands if they need help or spot a burgeoning problem.... Red flags should not evoke punishment but encouraged as they identify concerns early enough for senior leaders to intervene as needed to address those concerns."
> Source: *Strategic Initiative Management: the PMO Imperative*, Boston Consulting Group, Project Management Institute, November 2013.

Self-Assessment Checklist

The following self-assessment assists the reader in identifying strengths and opportunities for improvement against the key performance dimension that we consider critical for succeeding with strategy management in the digital age.

For each question, any degree of agreement to the statement closer to one represents a significant opportunity for improvement (Table 8.2).

Table 8.2 Self-assessment checklist

Please tick the number that is the closest to the statement with which you agree		
	7 6 5 4 3 2 1	
My organization is generally very good at managing strategic initiatives		My organization is generally very poor at managing strategic initiatives
In my organization, the strategy office and PMO work very well together		In my organization, the strategy office and PMO do not work well together
In my organization, strategic initiatives are generally owned by senior managers		In my organization, strategic initiatives are general owned by mid-level managers
We manage strategic initiatives as an integrated portfolio		We manage strategic initiatives independently for each other
We are very good at measuring the performance impact of initiatives		We are very poor at measuring the performance impact of initiatives

References

1. Jim Highsmith, et al. *Manifesto for Agile Software Development*, Agile Alliance, February 2001.
2. Rafael Lemaitre, James Creelman, Roberto Wyszkowski, *Creating the Transformation Office, A New Organisational Capability for the Digital Era*, ShiftIn White Paper, 2016.
3. Abridged from Palladium's 2013 Hall of Fame Report and Other Materials.

9

Unleashing the Power of Analytics for Strategic Learning and Adapting

Introduction

In Chap. 4: *Strategy Mapping in Disruptive Times*, we explained how advanced data analytics was finally enabling the realization of the "promise" of the Strategy Map. That promise being the understanding of the causal relationships between the objectives. Hitherto, causality was suggested through placing "arrows," on the map, but with little (mostly zero) testing of those assumptions (Fig. 9.1).

Indeed, the arrows often made the Strategy Map resemble, as Peter Ryan (a two-time inductee into Kaplan and Norton's Balanced Scorecard Hall of Fame and presently a senior executive with Christchurch City Council) notes, "a plate of spaghetti and meatballs."

Ryan continues that if arrows are used to show all the relationships between objectives on the map then it *has to* look like the popular Italian dish, simply because, "there are hundreds of ways one objective impacts another, both vertically and horizontally as well as across themes."

This is an important observation. As we also stated in Chap. 4, how value is created is as much Quantum and Newtonian mechanics. Pulling a lever in one place (to improve performance to an objective) does not necessarily lead to an automatic effect elsewhere (on another objective). Causality, as statisticians repeatedly tell us, is very difficult to prove. Furthermore, they remind us that correlation does *not* equal causality: perhaps the first law of statistics.

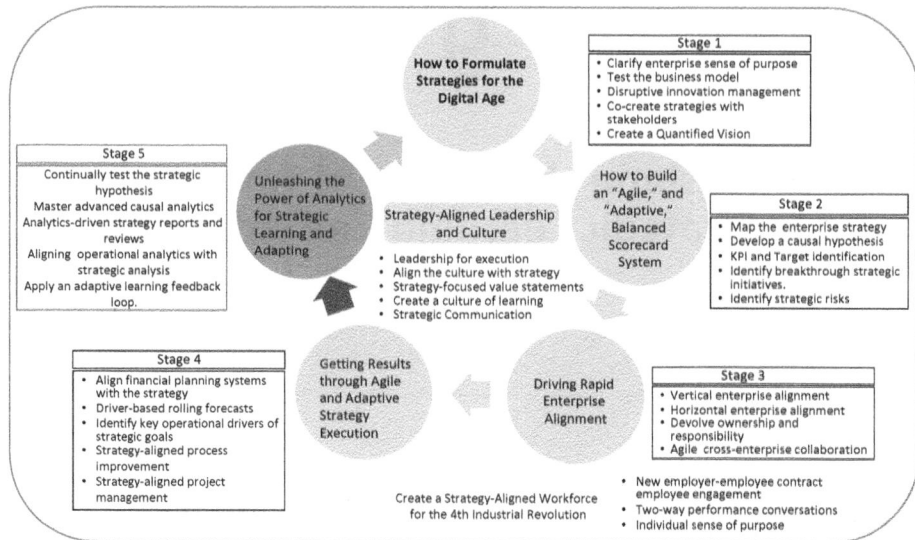

Fig. 9.1 Stage 9. Unleashing the power of analytics for strategic learning and adapting

Correlations and Causality

Here's an amusing story. A large organization hired a consulting colleague of one of the authors (and a recognized global thought leader in data analytics) to answer a very curious and perplexing question. Why had their stock price suddenly shot upwards on several occasions over the previous few years, without any apparent reason?

Therefore, the consultant ran some analytics and discovered something interesting. On the exact same day—over the timespan analysed—that the stock spiked, the lawns at corporate Headquarters (HQ) were mown. The data detected a direct correlation. Now what was the link? Did mowing the lawn affect the behaviour of investors? Of course not. The analyst looked at whether important events took place at the HQ on those days, or soon afterwards. No. The stock price spike and mowing the lawn were pure coincidence; there were other variables.

This might seem obvious, but now consider if such correlations were found between variables that seem logical.

For instance (and another true story), a marketing campaign was launched to reduce the number of customers defecting to the competition. To keep the customer loyal, the campaign offered discounts and other incentives.

At the end of the campaign it had "worked"—defection rates had fallen. So causal assumption: this marketing campaign resulted in the desired effect. Proven.

Well, no. Later analysis found that, at the same time that the campaign was in play, the main competitor had experienced a major operations issues and had scaled back efforts to recruit new customers (especially from the competitor). This analysis found that the discounts, and so on, had very little effect on the customers' behaviour. When the competitor scaled back up, the customers began defecting and at the original high rate.

The moral of these two stories is that data correlations do not necessarily prove anything and should always be tested against other evidence, other variables, and sometimes common sense.

Analytics and KPIs

Indeed, the present rush to use advanced data analytical tools is akin to the rush, which started a couple of decades back, to get Key Performance Indicators (KPIs). Leaders assumed the KPIs would provide all the answers. Find the "right" KPIs and all problems are solved. Again, nope! Neither will answers automatically come out of advanced data analytics. Both are powerful aids for decision making. They do not, and cannot, *make* the decisions. And both might, and often will, provide misleading information.

As with KPIs (explained in Chap. 5: *How to Build an agile and adaptive Balanced Scorecard*), value will only be secured when the basics of how analytics work is understood. Organizations are awash with KPIs but rarely invest in providing basic training in how measures work. Care should be taken to ensure the same mistakes aren't made with regard to data analytics.

Analytics and Decision Making

With cautionary tales and words as a backdrop, we can begin to consider how analytics can indeed support more agile and adaptive decision making. And the power of data analytics is considerable. As Dr. Norton has commented, "advanced data analytics represents the next stage of the evolution of the Balanced Scorecard system."

A Data-Driven World

However, to put all of this into context, consider the following. By 2020, it is predicted that the data universe will have reached 40 Zettabytes. To put this mind-boggling figure into meaningful terms, if 707 trillion copies of the more than 2000-page US Patient Protection and Affordable Care Act were stacked end-to-end and stretched from Earth to Pluto and back 16 times, that would equal about one Zettabyte. Now multiply that by 40! That's a large volume of data and represents a 50-fold growth from the beginning of 2010. However, despite the ever-increasing volume, studies find that less than 3% of useful data is analysed [1].

As a further illustration of how we have entered a data-fuelled world, from January 2016–June 2017, we created more data than in all prior human history combined. More hours of YouTube videos are uploaded every three minutes than all that the many Hollywood studios produce in a whole year! We could go on.

Big Data Analytics

Improving from this small percentage will be increasingly aided through Big Data analytics—which provides the capabilities to analyse large volumes of data. Big Data can originate from anywhere, such as sensors designed to collect climate data, social networking sites, digital videos and images, cell phone GPS signals, and sale transaction records, among others. Big Data analytics explores concealed patterns and unidentified connections and provides other valuable insights into the data.

In the Big Data world, we speak of the 4Vs (Volume, Velocity, Variety, and Veracity), as explained in Fig. 9.2.

A Measurement Game Changer

Returning to the theme of KPIs, by implication, KPIs are backward looking (in that they provide insights into what has happened). Although long

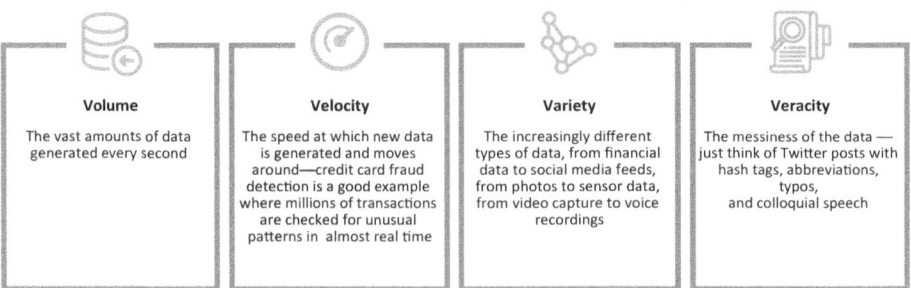

Fig. 9.2 The 4Vs of big data

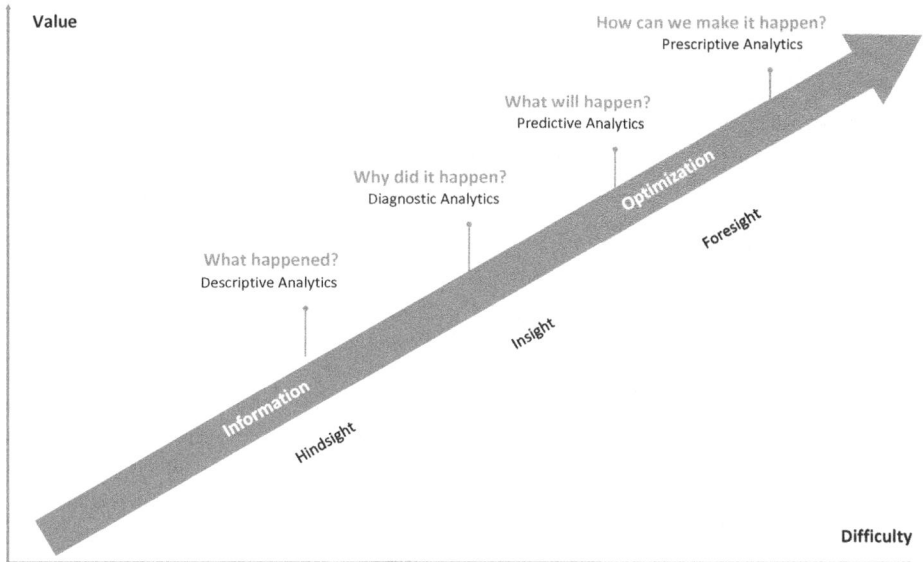

Fig. 9.3 Gartner analytic ascendancy model. (Source: Gartner)

recognized as having great value for decision-making purposes, the exponential expansion of data volume, velocity, and variety, and the ability to make sense of this messiness through Big Data analytics is a measurement game changer. The development of sophisticated statistical models to apply *Descriptive, Diagnostic*, and *Predictive* and *Prescriptive* Analytics is fast changing how we understand and measure performance. Figure 9.3 describes how Gartner view this as a model of ascendancy from descriptive to prescriptive. We will now explain each of the four stagers in turn.

Descriptive Analytics

Descriptive analytics identifies cause-effect relationships (or strong correlations) between data points, and now enables managers to answer the question, "What happened and why did it happen?" more confidently than previously. Historically, the "what" and "why" have largely been through gut instinct or experience.

Thus far, advanced descriptive analytics has largely been applied within operations to analyse the cause-effect relationships of driver and output metrics at the process level. Yet decisions and planning are still largely based on manual analyst estimates.

Diagnostic Analytics

Diagnostic analytics examines data or content to answer the question "Why did it happen?" and is characterized by techniques such as drill-down, data discovery, data mining, and correlations.

Predictive Analytics

As a next step, some organizations use analytics to make quantified and automated estimates about the future, including quantifying risks, based on the current and historical data and the cause-effect models identified through descriptive analytics; this is known as Predictive Analytics.

Predictive analytics automates forecasting, thus allowing the forecaster to test many different hypotheses in a short period of time, as opposed to the conventional manual forecasting approach, which is time consuming and does not allow for multiple hypothesis testing. However, to date, the focus of the use of big data and analytics has been within the commercial operations of organizations and increasingly within support functions such as finance, human resources (HR), and procurement.

Prescriptive Analytics

Prescriptive analytics not only anticipates what will happen and when it will happen, but also *why* it will happen. Further, prescriptive analytics suggests decision options on how to take advantage of a future opportunity or mitigate a future risk and shows the implication of each decision option.

Advanced Analytics and Strategy Management

From field observations, it is evident that there is little use for advanced analytics within strategy departments (often called an Office of Strategy Management—OSM).

Use of data within OSMs has typically been limited to the use of Business Intelligence software to extract and monitor the KPI performance in the form of Balanced Scorecard reporting systems. Since descriptive and predictive analytics were mainly used in operational decision making, the OSM was isolated from this development and thus never developed the analytic skills and capabilities. It therefore has not developed ways to integrate these advanced analytic aids into its own internal processes and work flow. As a result, advanced analytics has played a marginal role in strategy management. Most of the tools used for scorecard management are little more than reporting mechanisms—causal analytics is missing.

However, advanced data analytics can be applied, and with remarkable benefits, at the Strategy Map and Balanced Scorecard level and, as we now explain, at each stage of the Agile and Adaptive Strategy Execution model.

Develop and Translate the Strategy

The most obvious role that advanced analytics can play is at *Stage 1, How to Formulate Strategies for the Digital Age* and Stage 2, *How to Build an agile and adaptive Balanced Scorecard System.*

A strategy, graphically translated into a Strategy Map, is a collection of objectives that relate to each other through cause-effect relationships. An organization needs to achieve certain learning and growth and internal process (enabler) objectives that, so goes the hypothesis, will drive its external (outcome) customer/financial objectives.

Traditionally, these cause-effect hypotheses were based on in-depth analysis of the internal and external environment, along with the informed opinions of the in-house team and external consultants working on the strategy formulation. These hypotheses were then quantified in a financial model, but the driver assumptions and their cause-effect relationships were difficult to quantify accurately due to the qualitative nature of the external and internal analysis.

With descriptive analytics, organizations are able to use the rich data volume, velocity, and variety to quantify the cause-effect relationships between the strategic objectives. From this analysis, they can translate these qualitative cause-effect links between objectives into quantified formulas based on historical data.

The next step is to use predictive analytics to quantify what the future could look like if the organization achieves its enabler objectives. It can also conduct scenarios and assess risks of what the outcome objectives could look like if there is a *failure* to achieve. With quantified cause-effect formulas, it is now possible to define the objective targets for the enabler objectives at the *exact* levels required to achieve targets for the outcome KPIs and objectives.

Align the Organization

During organization alignment and strategy cascading (*Stage 3 Driving Rapid Enterprise Alignment*) a similar approach can be adopted for the business unit and functional strategies. As these cascaded maps also have cause-effect relationships, the use of descriptive and predictive analytics can play the same role as the corporate level.

Align Operations

As part of aligning operations (*Stage 4, Getting Results through Agile Strategy Execution*), an organization needs to define and set priorities and targets for the driver processes for each strategic objective. Descriptive analytics can be used to quantify the incremental impact of each percentage improvement in the driver process metrics on the percentage improvement of the linked strategic objective. The quantified cause-effect relationship between the driver process metrics and the objectives ensure that the targets set for the operational dashboards are in fact those that ultimately help achieve the targets of the linked objectives.

Strategic Learning

One of the key tasks of strategy reviews (a key part of stage 5, which we are considering in this chapter) is analysing root causes of under-performance within the objectives. OSMs frequently make assumptions about the root causes by using industry expertise within the organization.

These assumptions and theories are valuable, but should then be tested by the OSM using descriptive analytics to probe whether the numbers support the hypothesis. Predictive analytics can then be applied to project how the scorecard of the organization or function should look by the end of the next review cycle, if all the underlying drivers progress at current speed. Scenarios can be developed to model how performance could evolve under different circumstances and thus aid in prioritizing the action plan developed at the end of each quarterly review.

Strategy Update

As part of the strategy update (also a key component of stage 5), the organization can, using the historical data since the last strategy formulation or update, test the validity and accuracy of its cause-effect hypothesis of the strategy using descriptive analytics. They can then fine-tune the quantified formula cause-effect relationships between the objectives so that the financial model reflects more accurate quantified relationships and the targets for cause-effect related objectives are aligned (i.e. achieving targets for a driver objective, in fact, does result in achieving the target for the affected objective).

As a case illustration, Panel 1 provides an example of how descriptive and predictive analytics have been used to identify the root cause of strategic objective under-performance of a major Gulf-based telecommunications operator.

Customer Engagement

Consider too how powerful insights from advanced data analytics can improve customer engagement (typically key to any Strategy Map and scorecard).

Think of retail organizations selling to consumers. Such organizations can get to know their customers better based on their past relationship with the retailer (e.g. sales habits, customer care history), alongside a variety of digital footprints left by customers online: conversations on social networks, visits to various sites, and so on.

These organizations can cross-reference information in real time with customers' locations and inventory data at nearby shops. They can then send real-time recommendations to customers, offering them specific products at the nearest branch as they pass by. Walmart, as one example, sends personalized location-based coupons to your mobile phone based on such logic, enriched with additional data, such as the current weather (to ensure not to sell a barbecue grill when it rains).

As a further example, KLM Airlines identified passengers that were about to fly and tracked their behaviour. Based on real-time information from check-in records or Foursquare, they were able to surprise some of these passengers, giving them a small gift at the airport: for example, an iTunes voucher for a passenger who tweeted that s/he just bought a new iPad [2].

Required Capabilities

Naturally, for a strategy office to perform such advanced analytics requires these skills to be present. From our experience, we strongly recommend that advanced analytics should be a prerequisite for a strategy office (OSM) to be effective in the digital age. However, depending on the structure of the organization, these skills might be either within the OSM itself or in other departments, but utilized by the strategy department. What matters is that powerful, accurate, and data-driven decision support becomes the norm and that the OSM has the skills to present these findings.

Leadership Commitment

The leadership/management team must also value, understand, and sponsor the use of data-driven decision making and increase its reliance on analytics across all steps of the strategy execution process. For some senior managers, this means sacrificing much (but not all) of their career reliance on gut instinct and experience.

Data Management

Moreover, data has to be optimally managed. Most organizations capture a lot of data, but it is dispersed across siloed systems, thus making it a huge challenge for analytics teams to conduct analysis and correlation across the various data points. There several key steps to overcoming this challenge:

- Capturing all relevant data points—the transaction level detail.
- Ensuring the data is clean and validated (frequently data entered into systems have tons of errors).
- Integrating the data across different systems and sources (including external data sources such as surveys or census data) so that they can be correlated with each other (e.g. using the customer ID to link the data of the customer from the call centre system and from the billing system).
- Providing the necessary analytics tools that allow the expert OSM members' teams to interact with the data.

> **Advice Snippet: Implications of Big Data for Management**
>
> The ability to analyse and make sense of big data is an extension of the KPIs we use to monitor strategic performance and identify required initiatives or process improvements.
>
> Those organizations that have learned to truly leverage big data understand that they must analyse their new sources of information within the context of the bigger picture; in other words, integrating Big Data with traditional data to gain a 360-degree view of the extended enterprise.

> Big Data Analytics is enabling organizations to convert their performance data into relevant information and knowledge, allowing management teams to hold richer and more informed performance conversations and make more evidence-based decisions.
>
> Organizations can pinpoint the specific components of a KPI that really make a difference and ensure that actions and initiatives are driving performance improvement against that element.
>
> Moreover, based on good data analytics, OSMs can better advise management teams that, "this is where you really need to focus your attention right now."

Simple Analytics

However, useful analytics does not necessarily have to be advanced. Simpler analytic techniques can lead to the uncovering of insightful learnings. Many well-established tools, such as Fishbone, or Ishikawa Diagrams, named after the inventor, Fig. 9.4, are used to determine the root cause(s) for not achieving a desired outcome (effect). This is approached from the negative "what would cause us to not be able to provide our customers convenient access to the right products."

A tree diagrams is another useful tool for analysing layers of causes, looking in-depth for the *root cause*. Often called Five Ways, it starts with the effect and the major groups of causes and then asks for each branch, "Why is this happening? What is causing this?" Fig. 9.5.

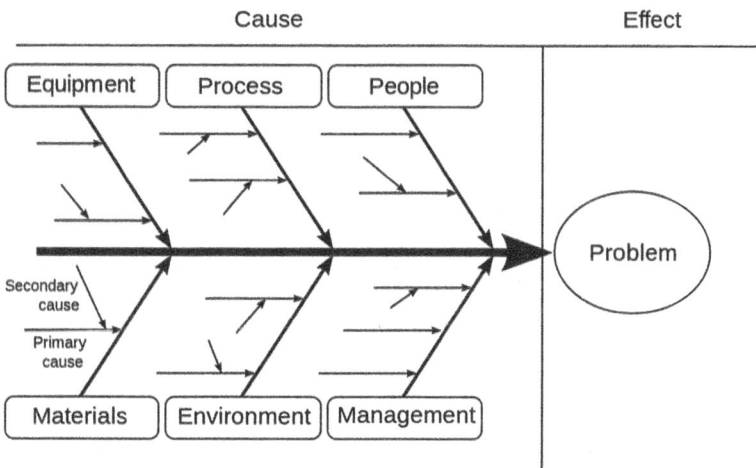

Fig. 9.4 A Fishbone diagram

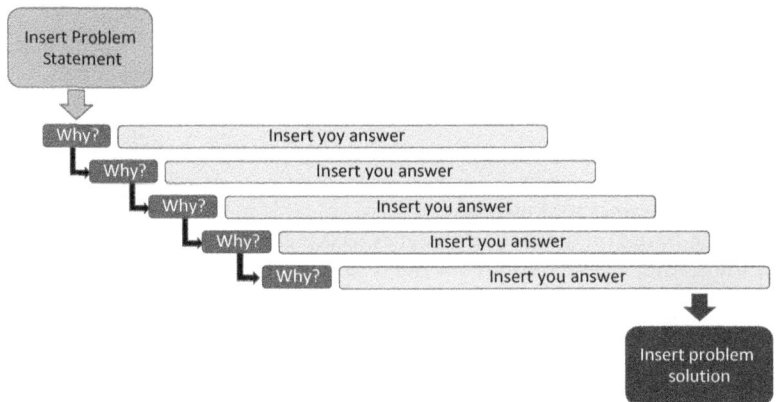

Fig. 9.5 The five whys problem solving tool

Doing simple analysis of data on, for example, a customer experience survey can also yield valuable insights—this reinforces a key message of Chap. 4 that the high-level KPI "score does not tell the whole story." Interrogating the underlying data is required to get a more insightful view of performance.

UK-Based Airline Case Illustration

For example, and as another airline story, a UK-based carrier surveyed its customers and found that 80% rated their overall satisfaction score as 4/5 in a 1–5 score (5 being the top rating). Consequently, the colour of the Customer Excellence objective was Green.

But, the score of 80% does not provide any useful information—that's where analytics comes in. Each customer was asked to provide two ratings. One for their actual experience against each of six dimensions (check-in, being kept informed, friendliness of staff, airline comfort, airline food, and flight timeliness) and the other for its importance to them.

Simple analytics found that for most customers the overall rating was based on how they experienced those components that were the most important to them (not an average of all). For the majority of customers, the most important areas were "being kept informed" and "friendliness of staff."

As shown in Table 9.1, a breakdown of the component scores for those who provided an overall rating of 5 also scored 5 for "being kept informed" and "friendliness of staff" and less for certain other areas.

Table 9.2 shows that who provide an overall rating of 3 provided an experience score of 3 for these two areas and importance of 5. They rated other dimensions as 5 for experience, but less for importance.

Table 9.1 Performance area breakdown for customers that provided an overall rating of 5

	Check-in	Being kept informed	Friendliness of staff	Airline comfort	Airline food	Timeliness
Importance to customer	4	5	5	4	4	4
Experience score	4	5	5	5	4	5

Table 9.2 Performance area breakdown for customers that provided an overall rating of 3

	Check-in	Being kept informed	Friendliness of staff	Airline comfort	Airline food	Timeliness
Importance to customer	3	5	5	4	4	4
Experience score	5	3	3	4	5	5

So, based on this simple analysis the airline knew which performance areas it had to excel within to provide the experience the customer wanted. By focusing attention and investments into these "emotional touchpoints," (see also Chap. 2: *From Industrial to Digital-Age-based Strategies*) while ensuring the other areas were provided at a good level and to industry standards, they could expect overall experience scores to improve. The 80% score now had meaning.

Performance Reviews

In the strategy management process, analytics are key inputs into the regular strategy review meetings. According to Kaplan and Norton, there are three types of review meetings.

1. Operational reviews: with the goal to respond to short-term problems and promote continuous improvement and informed by operational dashboards (often held weekly, or sometimes daily)
2. Strategy Reviews: the goal being to fine-tune the strategy and make mid-course adaptations. Informed by Strategy Map and Balanced Scorecard reports (often held quarterly, but sometimes monthly, especially applicable for particularly fast-moving markets)

3. Strategy Refresh: the goal of improving or transforming the strategy and informed by the Strategy Map, Balanced Scorecard, profitability reports, analytic studies of strategic hypotheses, external and competitive analyses, and so on, (typically held annually).

We will consider each meeting in turn.

Operational Reviews

As well as covering wider day-to-day issues, the operational review should also be used to track those processes that have been identified as the key drivers of the strategic processes that are housed in the Strategy Map (as explained in Chap. 7: *Aligning the Financial and Operational Drivers of Strategic Success*). Any issues concerning strategically critical operational processes can be escalated to the strategy review meeting.

Strategy Reviews

The strategy review is, from our experience, often an opportunity wasted. Too often, it is a lengthy discussion about the colours on the map and therefore the focus is squarely on the performance of "red" KPIs. Such events feel more like a "name and shame," show-trial than a forum for strategic conversations. Few leadership teams think about the overall performance of the map (which is much more than the measures) or the impact of the initiatives. The sole worry is whether this quarter's KPIs will be red, amber, or green.

The Shortcomings of Colour Coding

Colour coding is now, and had always been, a dumb idea. No matter how hard we try to persuade otherwise, most people want to avoid being red—it is after all the universal symbol of danger. In the workplace, red is a symbol of poor performance and therefore possible punishment and even termination. Avoid at all costs. And avoid people do—from fighting for easier targets or by manipulating the data.

This problem is further compounded by the fact that accountability is usually assigned according to functions, so the objective is perceived as a reflection of the performance of that function and by implication the ability of the manager. One of the authors was once told by a functional head that he would not accept a KPI target because he was worried it might be red, and if so, he'd be fired due to his function under-performing. Not an uncommon response.

Indeed, the very premise of cause and effect makes nonsense of assigning individual accountability. Objectives (and their supporting KPIs) are interdependent. No one objective delivers value in isolation and cannot be achieved without the input of others. Simply put, a red or green colour cannot be the result of a single person's/function's efforts.

Consider outcome objectives. People can be given accountability for a financial or customer objective, but these are simply outcomes of the work done in the enabler perspectives (internal process and learning and growth). So, can a manager with say accountability for a Customer Experience objective be held accountable for the achievements of the managers of the driving enabler objectives?

Similarly, an owner of an internal process objective is dependent on the capabilities delivered by learning and growth objectives and oftentimes the performance of other internal process objectives. Similarly, the owner of a people-related objective such as "High Performing Culture" and supporting KPIs such as "Employee Engagement" (typically from HR) cannot possibly deliver to these without significant work by leadership and other managers.

Accountability at the Theme Level

This is one reason why strategic themes (see Chap. 4) are very important on a Strategy Map. It is a recognition of causal flows and inter-dependency between scorecard system elements. Accountability can be assigned to theme team leaders, who report the collective performance of that theme, but overall accountability should rest with the complete leadership team. Strategy is a team sport, not a contest between individuals. Think soccer rather than golf.

Case Illustration: KiwiBank

Palladium Hall of Fame inductee KiwiBank provides this illustration of the value of strategic themes in better assigning accountability. When first introducing the scorecard in about 2008, an early learning was that assigning accountability to objective owners simply perpetuated an already problematic silo mentality. As a result, the organization created cross-functional theme teams structured around the four themes of Excellence in Business Processes, Sales and Service Leadership, Sustainable Growth, and, interestingly, Learning and

Growth (the reason being that learning and growth objectives work together to drive value at the internal process level).

Every executive officer was assigned to at least one enterprise-level theme team. About 150 employees (17% of the staff) were enlisted as team members, business unit strategy leads, objective champions, and measurement leads. In this way, all the work was orchestrated by the theme owner and, although responsibilities and reporting structures were still in place, the focus was on collective accountability and high-quality performance conversations, rather than silo-based objective management.

Strategy Reports

The goal of ensuring the strategy review is a dialogue rather than simple KPI reporting is somewhat stymied by the propensity of strategy office's (OSMs) to develop overly detailed strategy reports. We have seen reports that literally run to hundreds of pages, with a supposed "analysis" of each KPI—analysis that is little more than a collection of updates from each responsible manager, who often resent this quarterly intrusion into (and assessment of) the work of their function. Of course, the reports are rarely, if ever, read. Such an approach is unsurprising perhaps, as the KPIs generally provided the sole insights into how the objectives are performing. This is about to change and will have far-reaching implications for the strategy review.

A Decline in the Importance of KPIs

Although KPIs will remain important, we foresee that in the next evolution of the Balanced Scorecard System measures will remain a key focus at the outcome level (financial and customer), as they capture what has happened, but will be seen more as a snapshot in time. A fleeting moment of performance outcomes in a world where changing customer needs, increased competition, and continuous disruption will be the only constants to rely on. Being comfortable with today's outcomes would become an act of madness. Nevertheless, the financials, which have evolved since a Venetian Monk invented Double-Entry Bookkeeping in 1492, will remain critically important for some time, we expect.

Moreover, organizations should integrate rolling financial forecasts into the quarterly review (see Chap. 7). When done properly, these forecasts provide an honest assessment of likely future financial performance and typically to 4–6 quarters ahead. See Panel 2 for a case illustration of using rolling forecasts alongside the Balanced Scorecard and reporting these in the quarterly review.

At the enabler level (internal process and learning and growth), KPIS will become less important. Already, we have noted that the complexities around inter-dependencies here make assigning individual accountability nonsensical.

Organizations will gradually let go of their obsession with finding the perfect strategic enabler measures and be more interested in analysing and acting on the data that informs what is driving the outcomes. Advanced analytics will be critical here. Yes, there will still be some KPIs, but used much more as guides, an input into the performance conversations rather than absolute measures of progress.

This is what we envision, the Strategy Map will grow in importance along with the actions and initiatives that drive change. Cause and effect, or more realistically the strength of correlations, will become the centre of attention within strategy review meetings. Reporting to objectives within the customer, process, and learning and growth perspectives (if we indeed retain these names) will be more about the headline finding. Think of how newspapers report. Main headline (Men Walk on the Moon) supported by a high-level narrative and graphical information.

More advanced scorecard systems will enable the drilling down to as much granular information as required by the user (the headline on page one supported by further details on pages 4 and 5 if the decision makers want to know more). And here, managers will likely find as many KPIs as they wish, which would have been automatically captured and calculated through online systems (so the KPI junkies are kept happy).

> **Advice Snippet**
>
> Periodically, when preparing the report for the strategy review meeting, print these words from Albert Einstein on the front cover. "Not everything that counts can be counted and not everything that can be counted counts." [3] A gentle reminder that the meeting should be a conversation on performance and not a discussion on KPIs, which are simply an input into the discussion.

The Strategy Refresh

The annualized strategy refresh, the third of Kaplan and Norton's meeting types, is increasingly criticized as being out-of-sync with the speed by which markets move. The external world will stand still under the next scheduled planning cycle and resulting strategy refresh, yet this is precisely how many organizations behave. Plan, freeze for a year, refresh.

Research Evidence

As we explained in Chap. 1, early in 2016, a research project led by The Leadership Forum Inc., asked more than 200 mid-level managers and their direct reports in a global B2B company a simple question: "what are the three factors that most inhibit the execution of strategy in your business unit or more focused segment of the business?"

What was startling about the replies was not what was said, but what was not. Although many words were applied to the internal barriers to strategy execution (inconsistent leadership, poor communication, lack of clarity, etc.), very little was said about the external barriers. Seemingly, little thought was given on a day-to-day basis as to how marketplaces change, especially how the actions of rivals, customers, and other actors could overwhelm their plans.

The Price of External Failure

The almost complete absence of concern with external factors unearthed by the research meant that:

1. *Focus was almost exclusively internally oriented.*
    In the absence of an embedded external focus, the ongoing competitive dynamics will no doubt alter the underlying marketplace opportunity—for better or for worse—and do so before it is anticipated or detected.
2. *Capability centred almost entirely on internal capacities.*
    In the absence of attention to marketplace-oriented capability enhancement, internal capabilities may be augmented, but there is no guarantee they will contribute to a winning marketplace strategy.
3. *Coherence seemed to obsess on the internal relationships and linkages.*
    Coherence that is obsessed with an integrated plan misses the necessary balance between the internal and the external; it can quickly degenerate into reordering the deck chairs on the titanic, [4].

Learning and the Annual Refresh

What does this mean for the annual refresh? Simply, it means it becomes less important. It might still take place, but with a constant eye on external events, the strategy does not become essentially fixed for a year and simply about executing a plan.

That said, it would be chaotic, as well as organizationally and individually enervating if, for planning to be "everything," planners busily reworked plans daily based on challenges to assumptions. For the strategic assumptions to be effectively tested (and so, verified or not) requires that they be "frozen," for a time and with periodic "unfreezing," to analysis and discussion. In these events (which will typically be as part of the quarterly review, but that will likely increasingly take place on a monthly cycle), the quality of the conversations between the senior leadership and those executing in the field become critical. But this must start from a leadership mind-set that is focused on testing assumptions, rather than how effectively the strategy is being implemented. Equally, execution managers must also assume a mind-set that does not automatically default to blaming "the strategy" or other managers for execution failure. Governance is all about learning, rather than blame games or/and finger pointing.

Implications for the Strategy Office: Practitioner View

Chuck Biczak, Director, Strategic Change Management, Canon USA, observes that better use of analytics as well as the need for more real-time strategic conversations is fundamentally altering the role of the strategy office, or OSM. "To whatever extent any strategy office considers itself to be traditional, it is already outdated. The increasing speed of change within markets has resulted in increasing demands for an agile strategy office within companies," he says. "In the late 1990s and early 2000s, a 'good' strategy office may have worked with business units on articulating their strategies, identifying key performance indicators, and helping to manage strategic projects. Occasional reports to an executive team would close the loop on the performance of the strategy and integrate learnings into the next phase of the strategy, usually taking place within a pre-planned, annual strategy process."

He continues that over the last 10 years, strategic planning has come closer and closer to the businesses. "We don't only have strategy conversations with business units at pre-determined points of the year. Today, strategy conversations are the new norm and they are frequent and casual: in the cafeteria, in the hallways, and in almost every meeting. This greater integration of strategy into the day-to-day operations has not decreased the demands for a useful strategy office. In fact, it has increased the demands for a creative, responsive, and flexible strategy office that is sensitive both to emerging customer uses and the strengths of the specific organization" [5].

Parting Words

As we progress further into the so-called digital age, the use of advanced data analytics will become increasingly important and powered by computer-based tools (including machine learning—the application of artificial intelligence (AI) that provides systems the ability to automatically learn and improve from experience without being explicitly programmed). Such tools will be more and more sophisticated while also being simpler to use. Teams of highly skilled statisticians will be replaced by data analysts who will, in some instances, be highly qualified, while others will simply incorporate data analytical skills into their day-to-day work.

Those organizations that claim and maintain a competitive advantage will be those that have one eye looking at the internal world and one looking at the external—and utilizing, an everyday part of their work, the power of advanced data analytics to help spot sudden or emerging market/customer trends.

Scheduled meetings, such as operation and strategy reviews, will still take place and for good governance reasons, but will increasingly become more fluid and dynamic. The annual refresh will become less important and, for many companies, will simply disappear due to the pressures of continually synchronizing the internal and external rates of change. This has significant implications for strategic planners.

> **Panel 1: Example of Advanced Data Analytics Usage in a Telecommunications Firm**
>
> During the preparation of the quarterly corporate strategy review, the OSM team of a Gulf-based telecommunications provider identified significant underperformance in the customer perspective objectives related to stimulating revenues from the mass-market segment and winning a higher share of wallet from the high-value customer segment.

Applying Advanced Data Analytics

A first round of root-cause analysis did not identify an under-performance of existing internal process objectives as the root cause. However, the OSM team had developed a robust customer data mart that merged customer data from several IT systems into an analytical tool. Using this tool, the OSM's strategy analysts asked the OSM's data analysts (this OSM has analytics as a core OSM capability) to perform an analysis to understand what aspect of the customer behaviour was correlated with the under-performance of the customer perspective objectives.

The data analyst team identified a drop in on-network voice revenue as the key reason for the fall in the average revenue per user.

Further examination of the data indicated that while the used minutes had increased dramatically during the quarter, the average price per minute had dropped even more dramatically. Isolating the customers with a dramatic drop in price per minute indicated that they also experienced a dramatic increase in usage. Looking at the rest of the customer base, the data analysts discovered that a large pool of the other customers also saw a relative decline of their minutes of usage, but no drop in their price per minute, along with a relative increase in their incoming minutes.

To understand the reason behind this behavioural change, the OSM's data analysts examined the tariff packages, seasonal promotions, and campaigns adoption of these two types of customers.

By cross checking against the types of tariff packages and promotions that they subscribed to during the quarter, they found a common factor. The customers who experienced a dramatic increase in minutes and decrease of price per minute had almost all subscribed to an aggressive promotional offer launched at the beginning of the quarter by the Commercial Unit in response to a similarly aggressive offer by their main competitor.

The data analysts then discovered that the second pool of customers who had a drop in their minutes of usage but no drop in their price per minute were not subscribers to this aggressive promotion, but they were in fact receiving incoming calls from the other group of customers who were subscribers to the aggressive promotion.

The aggressive promotion was in fact an offer whereby customers were provided unlimited in-network calling in return for near doubling their monthly spend. The strategy analysts were able to conclude that the reason the customer perspective objectives were not met was due to the aggressive promotion, which was a tactical knee-jerk response to a competitor's offer. It was adopted by some customers who, in turn, were increasing their usage minutes calling other customers of the operator. These other customers, who did not subscribe to the aggressive promotion, were reducing their outgoing minutes' usage since the promotion-subscribed customers were calling them using their unlimited minutes.

The overall impact was that the promotion achieved its objective of increasing the spending of the target segments, but its side effect was that non-subscribers of this promotion had reduced outgoing calls as they were receiving incoming calls from the promotion-subscribed customers. The aggregate result across both customer segments was an overall drop in the revenues.

Strategy Review Reporting
Reporting this finding in the quarterly review to the management team, the OSM was able to bring to light the unintended consequences of the tactical response of the Commercial Unit to the competitor's offer. The management team then asked the Commercial Unit to find a solution to fix the problem and so stem the negative impact on the customer perspective of the corporate Strategy Map, [6].

Panel 2: Integrating Rolling Forecasts with the Balanced Scorecard: Case Illustration

In the first decade of the century, a Scandinavian financial services institution integrated the Balanced Scorecard, rolling forecasts, and strategic reviews and monthly reporting into one framework. It pulled together many of the elements of an integrated CPM solution.

The company's Balanced Scorecard comprised the four perspectives of financial, customer, internal process, and learning. These contribute to the delivery of the ultimate objective of "sustainable growth in economic profit." The Balanced Scorecard was the only process for annual target setting (it did not set an annual budget).

Targets were set during an annual strategic review, which included an analysis of economic, market, and competitive situation, with the appropriate revision to the group-level scorecard.

The targets were stretching and based on what was possible, not on what was forecasted. They were set at the top with each of its business areas having to shape plans to achieve them. The targets aligned to a three-year mid-term planning horizon, which forms a better linkage between longer-term goals and shorter-term strategic objectives.

The organization worked to a five-quarter rolling forecast, which was updated each quarter and based on the latest possible information from the business—including performance against the Balanced Scorecard, the impact of corrective actions taken in the previous quarters, and managers' assessment of the market and trading conditions. Forecasts were created at division, business area, and group levels.

The forecast was remarkably simple: an income statement including specifications tailored to business area specific characteristics. It typically comprised less than 10 line items that covered the critical drivers of revenues, costs and volumes. The forecast also included key performance indicators (KPIs) aligned to the KPIs in the financial perspective of the scorecard.

Monthly management meetings enabled costs to be analysed in greater detail than in quarterly meetings. The organization calls this "continual tactical performance monitoring."

Within the quarterly strategy review, business area leaders made presentations that started with the strategy report, which detailed performance of the map, KPIs, and strategic initiatives. This was followed by the updated rolling forecast that was based on data and commentary in the strategy report.

> "The quarterly meeting is a powerful forum for enabling a rich dialogue between senior management and business area leaders," said the Group Controller. "In it we are able to assess whether we have the right initiatives to meet our targets, test the cause-and-effect assumptions within the map and launch any corrective actions to close any performance gaps."

Self-Assessment Checklist

The following self-assessment assists the reader in identifying strengths and opportunities for improvement against the key performance dimension that we consider critical for succeeding with strategy management in the digital age.

For each question, any degree of agreement to the statement closer to one represents a significant opportunity for improvement (Table 9.3).

Table 9.3 Self-assessment checklist

Please tick the number that is the closest to the statement with which you agree	
7 6 5 4 3 2 1	
My organization has very good advanced data analytics capabilities	My organization has very poor advanced data analytics capabilities
The senior team fully understands that data analytics cannot provide all the answers	The senior team expects data analytics to provide all the answers
We are very good at using advanced data analytics alongside KPIs to gain performance insights	We are very poor at using advanced data analytics alongside KPIs to gain performance insights
We have very good data analytics capabilities within the strategy office	We have very poor data analytics capabilities within the strategy office
In my organization, strategic and operational reviews are very clearly separated	In my organization, strategic and operational reviews are not clearly separated
In my organization, strategy reports focus only on a critical few strategic priorities	In my organization, strategy reports are overly detailed

References

1. Webopedia Staff, *How Much Data is Out There?,* www.webopedia.com, 2014
2. James Creelman, Flora Lewin, *Big Data – Big Deal?* Palladium white paper, 2014
3. *Albert Einstein,* attrib.
4. Liam Fahey, Hubert St Onge, James Creelman. *External as well as Internal Perspectives… The New Strategy Execution Agenda.* Linkedin blog. October 1, 2016
5. Chuck Biczak, *Is the Conventional Strategy Office Outdated,* Strategically Speaking, Palladium, February 2015
6. James Creelman, Mohsen Malaki, *The XPP and Analytics: The Next Frontier in Strategy Execution,* Palladium white paper, 2016.

10

How to Ensure a Strategy-Aligned Leadership

Introduction

In the next two chapters, we explore leadership and culture. Based on experiences from many hundreds of scorecard implementations, we maintain that these are inextricably linked. The "shadow" the leader casts is long and powerful. Direct reports do what they know the leader wants, which is not necessarily the same as s/he says. This might not be found in corporate value statements, which we discuss in detail in the next chapter. We start with leadership (Fig. 10.1).

Perhaps exceeding strategy, innovation, disruption, process improvement, and so on, combined, countless billions (likely many trillions) of words have been written about leadership. What makes for a great leader? Are leaders born or made? and so on. Simply run a Google search for leadership quotations and you could be reading non-stop for the rest of your life (and still be nowhere near the end!). As for clichés, they are equally never-ending.

As with most other things, organizations look for the "magic solutions." Identify the characteristics of a great leader, hire, and plug and play. Sorted.

Alas, as with everything else in the management/organizational fields, the holy grail of leadership is a myth. Nevertheless, that does not (and should not) lead to our stopping the search for clues to its whereabouts. Useful clues are found in understanding context.

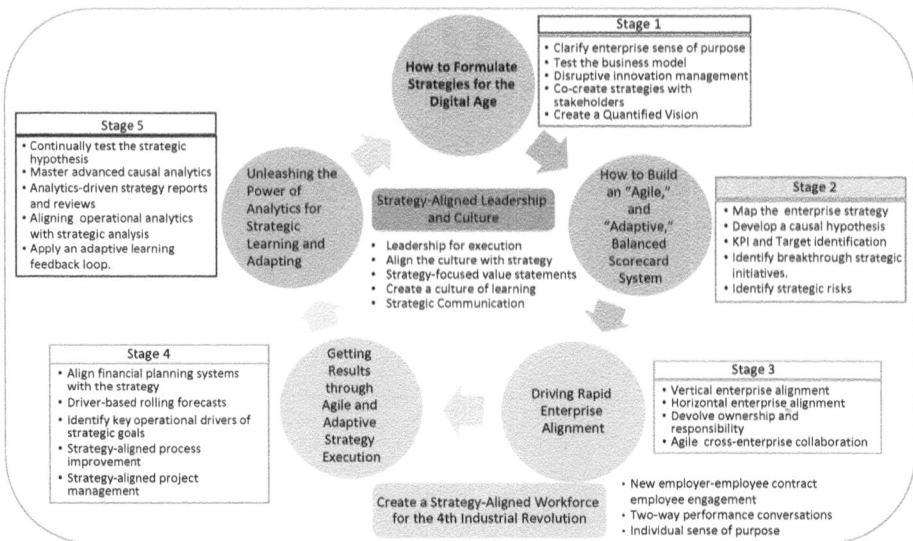

Fig. 10.1 Strategy-aligned leadership and culture

The Importance of Context

The Stories of Steve Jobs and Winston Churchill

Consider Steve Jobs. Was he a great leader? Well, many would automatically answer yes. A phenomenally successful leader of Apple. But, he was once fired (actually forced out—never technically fired) by the same organization in 1985. In retrospect, this might seem like a terrible idea (and may well have been), but consider some of the reasons for the dismissal.

Alan Deutschman explains in his book, *The Second Coming of Steve Jobs*, that first, the new version of Macintosh he had launched was failing to make money. Not in itself a reason to change the CEO, but he also had a reputation for, "[driving] people too hard. …being gentle and polite was not part of his demeanour."

He continues, "…he basically created his own team to create his own product, the Macintosh. His team had its own building. He even flew the pirate flag there. He said, 'It is better to be a pirate, than to be in the navy.' He had this company-within-a-company that became pitted against other parts of the company that actually made money." [1]

Jobs was out for failing to deliver financial results, being a bully, creating turf wars, and encouraging an "us and them" mentality in the same company. Few would argue against these as good reasons for dismissal.

Although Jobs later said that he learned from his mistakes, even when back in the driving seat at Apple and leading the development of industry disrupting and world-changing products (iPad, iPhone, etc.) simply transposing him to a mining company, working in a very stable sector, would likely have proven disastrous. Jobs' leadership style was appropriate for an organization reshaping the global communications landscape, but not appropriate for all industries. In 1985, it was deemed inappropriate for how the then board viewed the industrial context of Apple.

As a different example, consider Sir Winston Churchill. Certainly, another that is considered a great leader. His inspiring and rousing speeches ("we shall fight on the beaches," etc.) supporting a two-part strategy (make Great Britain a "fortress," before doing much actual fighting and then bringing the USA into the affray) were simply brilliant. Although notoriously difficult to work with, he had the right leadership style to deliver to his strategy.

But, in peacetime Britain, he was considered an ineffective leader (Prime Minister). He was simply too aggressive and did not reflect the, "calm and passive spirit that a peacetime leader should have. He was a constant reminder of war [and would constantly speak about the likelihood of another one]. He belonged in a war environment not peacetime Britain. It can be said that he was the reason for peace though he himself did not know how to maintain it" [2].

So, to the question, "what is strategy aligned-leadership?" what do we learn from these two examples? Leadership is, first and foremost, contextual. Different strategies require different leadership styles.

Leadership for the Execution of Strategy

Based on several years' rigorous research, The Palladium Group (for whom the authors of this book previously worked), in partnership with the Australia-based University of Queensland and Monash University, developed a "Leadership for the Execution of Strategy" (LFES) model [3]. As shown in Fig. 10.2, context is one of the 4Cs of the model, along with characteristics, conditions, and capabilities. Context provides the anchor for the other three Cs, as Jade Evans, a Palladium Managing Consultant and the lead researcher for Palladium and one of the architects of the model, explains,

> Without understanding the context that a leader is operating in, it is impossible to effectively execute strategy. Each organizational context is different, even if only slightly, and driven by a variety of factors, both internal and external. This means that without a clear understanding of this, and the ability to adapt to this context, the efforts to execute strategy will fall short. Our research showed that this is often where great leaders fail.

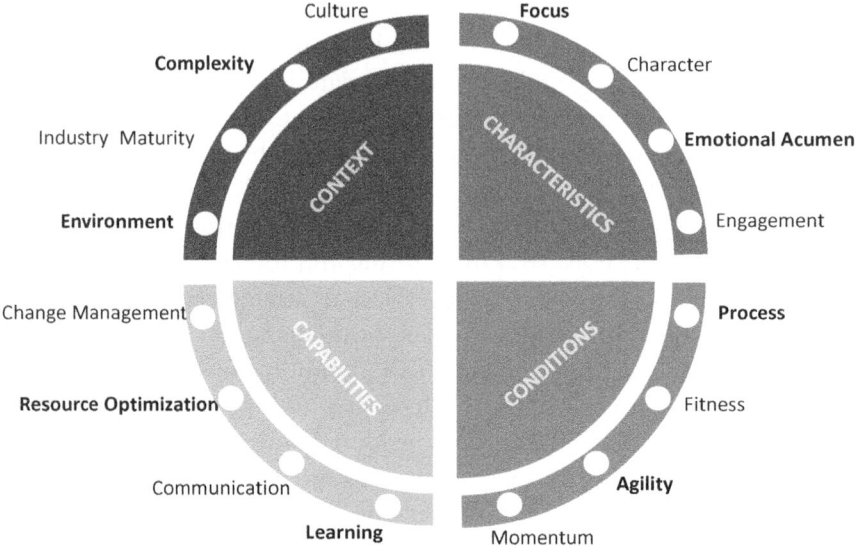

Fig. 10.2 The Leadership for the Execution of Strategy Model. (Source: Palladium)

Context

Palladium defines context as, *The internal and external enablers or constraints that accelerate or decelerate execution efforts in the organization.*

According to the LFES model, context comprises four sub-components (as do the other three Cs): culture, complexity, industry maturity, and environment.

- **Culture:** the history, narrative, values, style, business principles, and key cultural attributes of an organization can help or hinder leadership to drive successful execution. Strategy execution processes and capabilities must be built into the culture (we explore culture, specifically a strategy-aligned culture, within the next chapter).
- **Organizational complexity** is a significant factor in leader's ability to focus the organization. As a rule, leadership effectiveness is more critical as complexity increases.
- **The maturity** of the industry and the firm within that industry requires different leadership focus and different types of people. As organizations graduate through each stage of maturity, execution takes on different forms (as perhaps the Churchill examples shows—albeit at a national, governmental level).

- **The environmental** conditions (e.g. economic, political, etc.) play a key role in how leaders drive execution and may require different types of leadership and dynamic and adaptive strategy execution processes.

Having a clear understanding of the context and its implications sets the stage for leaders to struggle or thrive. Failing to do that can be catastrophic for an organization. Many companies collapse due to their lack of understanding of the context in which they are operating. Kodak is a good example of this. Their inability to respond to the emerging digital photographic market (which ironically, they invented) led to their loss of market share. Equally, Blockbuster's failure to see the emergence of online movies led to their eventual fall and their business failed as a result (for more on the reasons for the demise of these one-time industry giants, see Chap. 2: *From Industrial- to Digital-AGE-based strategies*).

Characteristics

The necessary distinguishing features that leaders in the organization must have for effective strategy execution.

The characteristics comprise focus, emotional acumen, character, and engagement.

- **Focus:** The ability of leaders to clearly focus and make decisions with either limited, often ambiguous, data or, just as likely in the digital age, too much data. Focused leaders know the right questions to ask, which details are important, and where their energy will deliver the best results.
- **Emotional Acumen:** Emotionally equipped leaders understand themselves and are adept at influencing and fostering challenge and stimulation in others. They connect with intellect and heart, bring people together, and manage consultative and directive behaviour.
- **Character:** Great leaders have strong character, moral fibre, and rigidly adhere to their core values. Character forms the basis of their integrity, and this drives their communication style.
- **Engagement:** The ability to motivate and engage employees and stakeholders to passionately grasp the vision for the organization is a key characteristic of all great leaders.

The characteristics of a leader are critical to their ability to execute an organization's strategy. Without the respect and admiration of the workforce, it is nearly impossible to take them with you.

More importantly, leaders who lack the characteristics required to lead successfully can make decisions that drive failure of organizations. Consider characteristics of the leadership team at Enron that led to a spectacular crash and an ensuing corporate governance scandal. While they were clearly focused on making the business succeed, they lacked the emotional acumen and character to understand the impact of their decisions and the engagement of their people to act as a moral compass for the business. This drove decisions that deceived the market, and ultimately caused the business to collapse and their investors to lose their money.

Conditions

The necessary organizational pre-conditions required for successful execution.

The leader's role in establishing the right conditions for strategic success comprises process, fitness, agility, and momentum.

- **Process**: Most leaders are unable to sustain execution success through sheer force of will alone. That is why having the correct processes is crucial. The right set of management processes and systems are required to enable alignment, drive accountability, and measure and manage performance.
- **Fitness** requires selecting the right people with the right mix of capabilities at the right place and time in the organization. Subtle changes in context can render a previously "fit" team "unfit."
- **Agile:** Organizations and leaders that successfully execute strategy over the long term must be agile and must be able to adjust their style to fit the culture, values, and business in which the organization operates.
- **Momentum:** Most organizations can sustain execution success in short bursts only, and are unable to sustain a continuous momentum over the long term. The best leaders create the conditions that position the business to succeed over the long term.

Leaders must create these conditions if the organization's strategy is to be successfully executed. Systematic failures are common, and the effects can be catastrophic. The global financial crisis is stark evidence of that. The lack of systems, checks, and balances was a key contributor to the collapse of the international banking system in 2008. The leaders of these organizations failed to ensure the right conditions were in place to execute their strategies successfully.

Capabilities

The critical capabilities that leaders at all levels must attain in order to sustain successful execution over the long run.

The components of the right capabilities are resource optimization, learning, communication, and managing change.

- **Resource Optimization:** The essence of execution is choice (as we explain in Chap. 2). Where, when, and how to allocate all types of capital: human, financial, time, and social/political. How these resources are optimized is key.
- **Learning:** Leaders are committed to learning; they are able to coach, mentor, and teach. More importantly, they choose people that have these same skills and create an environment of diversity, continuous improvement, learning, and innovation.
- **Communication:** The ability of leaders to communicate the vision and strategy consistently is the essential capability for leaders to impact and bring all levels of the organization into alignment. This includes tailoring the information and receiving feedback.
- **Managing Change:** Strategy execution is about managing change. Leaders that are able to convey the reasons for change, create a sense of urgency, and bring people along on the journey are the ones who are able to sustain execution success.

The Last Stage: Not the First

Note that managing change is the final of the 16 4C sub-components of the LFES model. This might at first be surprising, as this is precisely where many organizations start when executing strategy, especially when transformational. Evans explains, "Effective change management requires an intimate understanding of the pulse of the organization. Too many design and execute their change programs outside of this, limiting their effectiveness," she says. "What works brilliantly in one organization is rarely as effective in another. This is because each organization is different."

> **Advice Snippet**
>
> The Palladium Group's 4Cs LFES model is based on a maturity model, where each of the building blocks must be understood before the next can be obtained. Sequence is important.
>
> For example, issues around the internal and external context should be addressed so a leader can ensure that they are in fact leading with the strategic context understood. In turn, ensuring that the characteristics needed in an organization to effectively execute their strategies are in place is essential before analysing the conditions needed, such as having the right people—the right processes to execute is key.
>
> Until these three elements are in place—context, conditions, and characteristics—it is premature to work on developing capabilities.

Palladium Model Summary

The research that informed the LFES model provided valuable insights into how leaders work within organizations. The subsequent model helps leaders to communicate more effectively with the people within the organization, meaning that they can better target the change program. Further to this, it helps organizations understand how to use each leader, at each hierarchical level, to drive the change process, making it more bespoke to that organizational setting and the individuals within it, and therefore more likely to be successful.

The Potential of the "Ordinary" Leader

Perhaps a valuable answer to the earlier question, "are leaders born or made?" Evans comments that, from the research and supported by observations from work with hundreds of companies, Palladium came to understand that "ordinary" leaders could be equipped with the capability to successfully execute strategy through a comprehensive and focused process. "The research led us to reject the "hero leader" principle, whereby the rock stars of leadership provide a reliable model for leaders to emulate, though study of successful leaders can certainly be instructive," she says. "These 'hero leaders,' often savants in their own right (and therefore unrealistic to emulate), were often successful in a very specific context that cannot be replicated for other leaders who have unique challenges of their own."

Indeed, continually citing Jack Welch as the manager all should aspire to emulate is akin to asking all soccer players to try and play like Cristiano

Ronaldo (the Real Madrid star that many consider the best player in the world and one of the best of all time). Not realistic, however appealing.

Evans concludes that, "These findings clearly demonstrate the link between effective leadership and the ability to successfully execute strategy. More specifically, they show the need to develop a necessary understanding of context for the execution of strategy in organizational leaders, as traditional approaches to leadership development are not meeting the needs of organizations in the new economic realities."

She adds that traditional leadership development efforts in the majority of organizations do not explicitly address strategy execution. "Instead, leadership competencies that enhance general leadership effectiveness are developed, but they do not give due consideration to the strategic context of the organization."

Strategic Leadership: Research Evidence

Based on global research among almost 1300 organizations, the Palladium Group's 2014 Global State of Strategy and Leadership Survey (in which the authors of this work were involved, alongside Jade Evans) uncovered significant challenges for strategy leadership in the digital age [3].

Although 96% of respondents stated that strategic leadership was key to their future success, fully 51% found the quality of their organization's strategic leadership unsatisfactory. Alarmingly, 67% of board members, CEOs, and managing directors studied did not believe that their current leadership development approach was providing the skills that their leaders need to execute their strategies successfully, indicating a lack of executive stewardship of strategy execution, and even of the strategy itself.

Only 41% of respondents reported strong stewardship from the C-suite, without which strategy execution is typically de-prioritized. Participants identified managing change as the greatest weakness: 62% think that the inability to manage change or overcome internal resistance to change is a genuine problem within their organization. Fifty-six per cent say that strategy execution is challenging because it conflicts with existing power structures, a change issue incumbent upon senior leaders to resolve. This points to the performance sub-optimizing effects of the silo-based structure of organizations, as we stress throughout this book.

Poor Change Management

Leaders' poor change management skills are worrying. To paraphrase Dr. Norton, if leaders cannot manage change, they cannot manage strategy. Given the myriad changes that organizations currently face and the innumerable changes on the horizon, poor change management will have severe consequences for many organizations. In fact, of the organizations that identified change management as a major problem, more than 75% also claim to be poor at strategy execution.

This scenario is already playing out: the very organizations that are under threat in the market are the same that are worst at change (eight times more likely to find change management a major problem, compared to those whose value proposition is not under threat). Overall, the research found a high correlation between change management success and execution success. No organizations that excelled at change management performed poorly at strategy execution.

According to the research, only 26% of organizations align their leadership development to their execution needs—what Palladium calls leadership for execution. Leadership for execution, which includes creating a sense of urgency, making a compelling case for change, and committing the entire organization to the strategy, brings much higher returns, the survey found.

In terms of overall strategy execution performance, organizations that practice leadership for execution are 22.7 times more likely to be high performing (68% of high performers; 3% of low performers) against eight times for internal coaching, 8.7 times for structured leadership development programs, and 6.7 times for 360 feedback.

Leadership Development

The Palladium research also uncovered very worrying statistics regarding present approaches to leadership development. Fully 72% of companies did not think that their organization's learning and development programs are properly linked to the required skills for strategy execution.

Evans comments that, "If the budget repartition is '20% executive team/80% broader leadership team' instead of '80% executive team/20% broader leadership teams', an organization is three times more likely to develop the requested skills for strategy execution (12% instead of 4%)."

Summarizing Palladium's leadership development findings:

- Most leaders have no formal training in strategy execution.
- There is often a significant disconnect between leadership development frameworks and strategy execution.
- Leadership development programs in isolation do not adequately equip leaders with the tools required to lead strategic change.
- Leaders build up their ability to execute by making and correcting mistakes throughout their careers.

- Leadership is often unprepared for significant changes that disrupt the status quo.

Some very interesting learnings here for those involved in leadership development, be it HR organizations or consultancies.

> **Advice Snippet**
>
> Of all the important roles leaders play in strategy execution, leadership development is perhaps the most crucial to the long-term success of any organization. Developing the right leaders and building a culture and structure for them to stay with the organization and thrive is critical to any organization. It is also the biggest challenge that organizations face today and into the future.
>
> Leadership development solutions should be carefully tailored to the specific needs of each individual, team and organization instead of providing the usual "one-size-fits-all" approach to leadership development—the much sought-after perfect plug-and-play approach.
>
> Leadership for Execution should be the core steer in leadership development for digital-age organizations.

Agile Leadership in an Age of Digital Disruption

A further useful research project that speaks directly to the challenges of the digital age is the 2017 research by the Global Center for Digital Business Transformation. This analysed the requirement for "Agile Leadership in an Age of Digital Disruption" and identified key competencies and behaviours required to lead organizations in environments characterized by digital disruption [4].

According to this research, agile leaders are humble, adaptable, visionary, and engaged. "These must-have competencies inform their business-focused actions," the report's authors note.

Humble

Knowing what you don't know can be as valuable as knowing what you do in a time of rapid change, the research finds. "Agile leaders must be open, willing to learn, and seek input from both inside and outside their organizations. They also need to trust others who know more than they do."

Crucially, the authors state that they do not regard humility as an abdication of the need to provide a strong vision and positive direction, but as an acknowledgement that the current speed of change outstrips any leader's personal store of knowledge or experience. "Accepting that a single person cannot know everything needed to make a decision is a critical component of agile leadership," they write.

Adaptable

Being adaptable is simply an acceptance that change is constant. When new information triggers a leader to change their mind, it should be seen as a strength and not a weakness. "Focused adaptability based on new information is a distinct competency, unlike random vacillation," the report states. "Agile leaders adapt their behaviour in the short term based on their ability to make evidence-based decisions."

They add that, "… agile leaders are adept at dealing with complexity and less reluctant to change their minds in the face of new information. They are not afraid to commit to a new course of action when the situation warrants it."

Visionary

Visionary in the context of agile leadership means having a clear sense of the long-term direction, even in the face of short-term uncertainty. In times of rapid technology and business model disruption, with opportunities opening up everywhere, a clear vision is even more critical. "Agile leaders have a well-defined idea of where their organizations need to go, even if they do not know exactly how to get there."

However, the authors note that being visionary is not restricted to digital disruptors; it is equally important for legacy organizations, where leaders need to define and clearly articulate long-term aims and objectives. An example provided was General Electric's vision to become the dominant player within the "Industrial Internet," a term it coined, which is a major departure from the company's traditional manufacturing roots.

Engaged

Agile leaders display a willingness to listen, interact, and communicate with internal and external stakeholders and a strong interest and curiosity in emerging trends. In other words, they are engaged. "Whatever their hierarchical position, agile leaders are always engaged, be it with customers, partners, suppliers, team members, staff or the broader societal and industrial ecosystems. This desire to explore, discover, learn and discuss is as much a mind-set as a definable set of business-focused activities."

Essential Behaviours of Agile Leadership

The research also identified three key behaviours that differentiated agile from non-agile leaders: hyperawareness, informed decision making, and fast execution.

Hyperawareness

Hyperaware agile leaders are focused on spotting emerging digital opportunities or competitive threats. They are engaged, seek new insights, and adapt in response, but they are also aware of the need to provide guidance through a strong vision, as the potential for change threatens to overwhelm a linear strategy. "Good leaders are constantly scanning their environments, both inside and outside their organizational boundaries. With technology-driven change accelerating across industries, the need for leaders to look outward, and not just at their competitors, is evident."

Informed Decision Making

For agile leaders, informed decision making fundamentally entails using available data to make evidence-based decisions. To do this, leaders "… must recognize and utilize the best data sources, apply appropriate analytics, and then make a decision. Faced with insufficient or even contradictory data, leaders must draw on their experience and intuition to move forward."

Informed decision making underpins a leader's ability to adapt and support their long-term vision, the authors note. "Agile leaders who understand the value of using digital technologies to gather and analyse data are always on the lookout for new data sources to support informed decision making."

Fast Execution

Fast execution entails willingness on the part of a leader to move quickly, often valuing speed over perfection. "In an environment characterized by significant disruption, the effectiveness of hyperawareness and informed decision making is significantly reduced if the organization is not able to act with speed. Ultimately, agile leaders will only be effective if they are able to quickly execute an informed decision."

Assessing the Models

With differing spins, the research from Palladium and the Global Center for Digital Business Transformation show strong consistencies in their findings. As a few examples, both organizations point to the importance of the need for leaders to be adept at dealing with complexity and of being visionary in the context of having a clear sense of the long-term direction. Strong capabilities in execution are also highlighted by both.

Furthermore, both speak to the criticality of being open and willing to learn, to listen, interact, and communicate—perhaps, most importantly, of trusting others.

Our own research supports these findings. In the digital age, managing complexity, while ensuring a consistent message of intent and progress, will be leadership prerequisites. Moreover, conventional command and control structures, in which "managers think and workers do," must be seriously challenged and dismantled. No single leader can know all the answers and must make more use of the insights of a highly educated workforce (and we sense there is something wrong with that term) that has grown up with social media and so expect to learn, challenge, and debate.

Furthermore, we also strongly support two specific observations from the Global Center's research.

First, their assertion that good leaders constantly scan their environments, both inside and outside their organizational boundaries and not just their competitors. This supports a central message of this book that, in the digital age, we must move away from formulating a strategy and then attempting to execute, "as is," and only paying attention to internal obstacles to success. External environmental scanning must be an everyday activity and fully integrated into the strategy execution process.

Second, we also agree that an agile leader must recognize the value of finding and utilizing the best data sources and applying appropriate analytics to inform decisions. As we explain throughout this book, and most particularly in Chap. 9, Unleashing *the Power of Analytics for Strategic Learning and Adapting,* advanced data analytics is transforming the strategy management process (along with just about everything else). Analytics coupled with the insights of an increasingly technologically savvy employee-base will uncover significant opportunities for those organizations that figure out how to meld this into an agile and adaptive capability.

Parting Words

As stated at the start of this chapter, the holy grail of great leadership does not exist. Strategy-aligned leadership is very much contextual and, as with everything else, evolving. What is great in one environment might be a disaster elsewhere. Winston Churchill's leadership style was properly aligned to winning a complex war, but misaligned to the requirements of building a peaceful nation. Context matters.

But that said, there are a group of capabilities that, in some shape or form and applied with the proper understanding of context, is providing useful clues for good leadership in the digital age.

Interviewed for this book, Alistair Schneider, a Business Operations Leader and founder of startupsinnovation.com, succinctly summarizes many of the requirements of a leader in the early days of the 4th industrial revolution. "The type of leader required to drive transformative and continuous strategy execution in the Digital-Age will understand that the key elements are speed, agility, empowerment, presence, data analytics and acceptance of failure," he says. "Additionally, it is about making the mission the core principle that drives everyone's job everyday." The latter point supports our message in Chap. 2 that organizations must understand their sense of purpose. With that sense of permanence and consistency, then speed, agility, and so on will provide great value and the appropriate balancing of the longer term with the mid and shorter terms.

Anchored to context, in the final analysis, strategy-aligned leadership is essentially about execution. To conclude with one of the millions of quotes on leadership, we agree with the words of Ram Charan and Larry Bossidy from their seminal work *Execution: The Discipline of Getting Things Done*: "Execution is a specific set of behaviours and techniques that companies need to master in order to have competitive advantage. It's a discipline of its own" [5].

Self-Assessment Checklist

The following self-assessment assists the reader in identifying strengths and opportunities for improvement against the key performance dimension that we consider critical for succeeding with strategy management in the digital age.

For each question, any agreement to the statement closer to one represents a significant opportunity for improvement (Table 10.1).

Table 10.1 Self-assessment checklist

Please tick the number that is the closest to the statement with which you agree		
	7 6 5 4 3 2 1	
The prevailing leadership style in my organization is appropriate for our context		The prevailing leadership style in my organization is not appropriate for our context
My understanding has a very good understanding of what is required for the "leadership of strategy execution"		My understanding has a very poor understanding of what is required for the "leadership of strategy execution"
My organization has very good strategic leadership capabilities		My organization has very poor strategic leadership capabilities
My organization is very good at managing change		My organization is very poor at managing change
In my organization, learning and development programs are very well aligned to the required skills for strategy execution		In my organization, learning and development programs are very poorly aligned to the required skills for strategy execution

References

1. Alan Deutschman, *The Second Coming of Steve Jobs,* Random House, 2000.
2. Lauren Smith, *Winston Churchill: Why Was a Successful Wartime Leader Unsuccessful at leading a Peacetime Britain?* http://www.academia.edu
3. James Creelman, Jade Evans, Caroline Lamaison, Matt Tice, *2014 Global State of Strategy and Leadership Survey Report*, Palladium Group, 2014.
4. Michael R. Wade, Andrew Tarling and Remy El Assir. Contributor: Rainer Neubauer, *Agile Leadership in an Age of Digital Disruption,* Insights@IMD, May, 2017.
5. Ram Charan and Larry Bossidy, *Execution: The Discipline of Getting Things Done,* Crown Business, 2002.

11

How to Ensure a Strategy-Aligned Culture

Introduction

Of all the clichés that pepper the managerial lexicon, perhaps none rolls off the tongue more readily and regularly than Peter Drucker's, "culture eats strategy for breakfast." It seems that the authors of just about all the articles, blogs, and so on, that consider culture and strategy together are honour-bound to include this. It's as if it would be disrespectful not to (Fig. 11.1).

Yet, despite its frequent usage, the saying is still powerful and useful. It alludes to the truism that, whereas strategy formulation and planning are essentially straightforward processes or tasks that can be described, documented, and completed, executing strategies is something significantly much more complex.

From our observations, those strategy execution efforts that fail to deliver expected benefits do so not because the Strategy Maps and Balanced Scorecards are poorly designed (although many are) but due to a failure to adequately plan for, and then overcome, the myriad cultural bulwarks that litter the implementation highway.

Interviewed for this book, Andreas de Vries, an Oil & Gas strategy management specialist, believes that culture management should not be an afterthought to strategy. "The alignment of strategy and corporate culture should not be achieved by thinking about strategy first and corporate culture second, but by considering both topics together," he says.

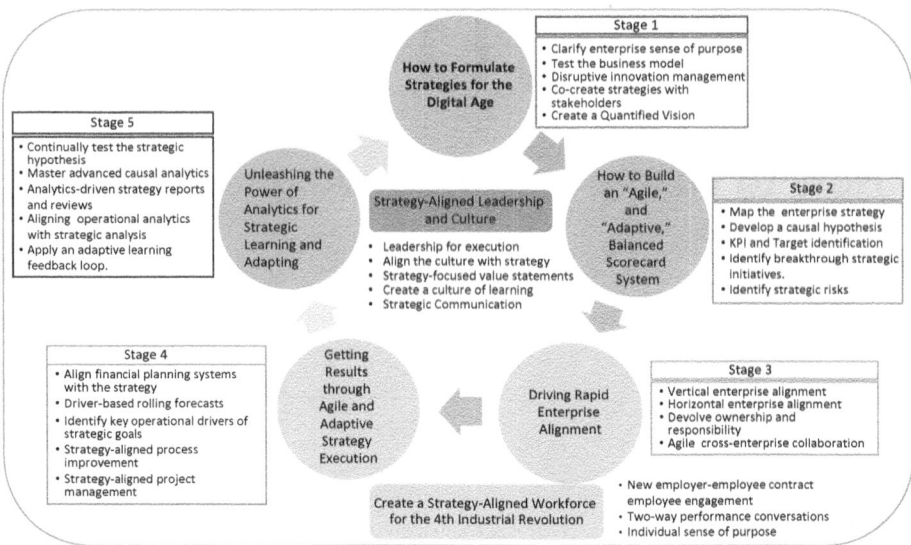

Fig. 11.1 Strategy-aligned leadership and culture

The Elephant in the Room

To use another cliché, organizational culture is the "elephant in the room." Big, powerful, and capable of trampling under its feet anything that it doesn't like, it is rarely addressed because, despite its size, it is virtually invisible. The extent of its influence, however, is not: untamed, dysfunctional, and unethical behaviour, bickering between co-workers, poor customer service, and employee disengagement from the goals of the organization are plain for all to see. By making the elephant visible, organizations have opportunities to shape behaviours and attitudes that support the strategic aims of the enterprise. The mighty elephant will be tamed.

The Challenge of Culture: Research Evidence

However, taming the beast is not an easy task. The Palladium group's 2014 Global State of Leadership and Strategy Survey report (based on almost 1300 replies from practitioners across the globe) found that fully 82% of executives believed their organization was not genuinely able to sustain a positive organizational culture. Yet, those organizations with a positive culture were 11 times more likely to attract and retain essential employees [1].

Defining Culture

To deal with the challenges, we must first define culture. An organizational culture can be described as a learned set of behaviours that employees recognize as those that are expected and rewarded. These behaviours are based on values that provide guidelines for decision making by defining what the organization praises and condemns. Values and behaviours are often built over many years, are deeply ingrained, and become "just the way we work around here." They are typically not easily found in HR policies or even well-crafted values statements (see below), but they are the most influential guides of "the unwritten rules of the game"—the way the organization really works—to which established employees pay attention and into which new recruits soon learn to assimilate.

Culture is the "mind" of the enterprise. Everything else is bodily parts and simply following instructions.

A Great Culture Is Not Necessarily a Strong Culture

A few words of caution: when looking to shape a strategy-aligned culture, the aim should not necessarily be to create a "strong" culture, which can be as performance constraining as one that is "weak." Think of Kodak (see also Chap. 2: *From Industrial to Digital-Age-based strategies*) clinging to "the way things are done around here" and recall IBM's strong technology-focused culture that eventually led to financial crisis before a new CEO drove a more customer- and sales-focused culture. As Harvard Business School professor Rosabeth Moss Kanter said in her book *Change Masters*, "Cultures will need to change over time as the tasks change, as the organization grows, or as people change. Much trouble in organizations comes from the attempt to go on doing things as they used to be done, from a reluctance to change the culture when it needs to be changed" [2].

De Vries supports this view, stating that a strong corporate culture does not automatically support strategy execution—it all depends on exactly which attitudes and behaviours the corporate culture promotes.

"If the promoted attitudes and behaviours are aligned with the strategy, it is a constructive corporate culture that supports achievement of the strategic objectives," he explains. "An example of this would be when a company that wants to enter or develop new markets encourages creativity, risk-taking and a customer-focus.

"On the other hand, if the corporate culture is not aligned with the strategy, it is a destructive or 'toxic' culture. An example of this would be when the same company that wants to enter or develop new markets encourages standardization, cautiousness and a cost-focus."

He argues that corporate culture can influence an organization's ability to execute strategy in four different ways.

1. If the corporate culture is weak and the alignment between the corporate culture and the strategy is weak, then chaos will characterize the organization. On the ground, different employees will display different ways of working, while the organization promotes yet another way of working that conflicts with the strategic objectives.
2. If the corporate culture is weak while the alignment between the corporate culture and the strategy is strong, the organization will be ineffective. The organization will promote a sensible way of working, but the employees have not bought into this way of working.
3. If the corporate culture is strong and appropriate but the alignment between the corporate culture and the strategy is weak, the culture will be obstructive. The corporate culture will influence the values, attitudes, and behaviours of the employees, but lead them to think, feel, and act in ways that go against the strategic objectives.
4. If the corporate culture is strong and appropriate while the alignment between the corporate culture and the strategy is also strong, the organization will be effective. In this case, the corporate culture is supportive of what the organization tries to achieve.

The point de Vries is making, and which we strongly agree with, is that in seeking to create a strategy-aligned culture, think about the type of strategy with which the culture must align. For example, if an organization is pursuing a strategy based on customer intimacy, then the defined behaviours, values, recruitment, training, and reward mechanisms, and so on, must be appropriate for the inculcation of customer-centricity, as should the structure, processes, information flows, and decision-rights. It holds true for a strategy based on operational excellence, product leadership, or whatever.

Corporate Values

The alignment of culture has many strands, but perhaps the most utilized are corporate values. However, the field observations of the authors (and this is agreed upon by just about every consultant and practitioner) find that value statements are too often "Motherhood and Apple Pie" statements. That said, there are organizations that prove otherwise (see the Statoil case illustration in Panel 1).

Typically, value statements sound good and no one argues with them—who would argue with a value around integrity? Thereafter they are hung on office walls and litter internal communications, but in practice are rarely implemented or adhered to.

Even less frequently do they lead to substantive change to organizational structures, especially around decision-rights. For instance, many organizations have values around trust, but how often is this supported by significantly redefining the monetary levels lower-level managers or employees can spend without signoff from higher levels? Or how many have loosened the strong micro-management of employee expenses? Not doing either sends a message to employees that they value trust, but they do not trust you. The unwritten rules of the game are strengthened. Policies, procedures, processes, information flows, decision making, metrics, and incentives must reinforce the new corporate values and desired behaviours. This is much more challenging than crafting the values and simply running awareness sessions to promote them.

Case Illustration: Poor Practice Example

A while back, one of the authors sat through a presentation on how one company was implementing a set of corporate values. The values were sensible (usual suspects—customer-focused, innovative, trust and empower) and the behaviours well described (and on a very nice poster, too). The author knew this company very well, so after the presentation (not during, so as not to embarrass anyone) asked the presenter how they were going to make the values come alive? The expected answers were forthcoming: awareness sessions for all staff, visual displays, coffee cups, and so on.

Based on the author's previous knowledge of this firm, he then asked, "Specifically about the trust and empower value, I assume you are now going to radically overhaul the rule by which lower level managers cannot spend any money without signoff from a senior manager?" After a pause, the presenter said there were no plans to alter this. Next question, "So I assume you are going to throw out the stupid policy that if an employee takes a day off sick they need a written note from a Doctor to prove it?" After a longer, and more embarrassed, pause, he admitted there were no plans to change this either. Finally, he was asked if they intended to abolish the clocking in and out system (which everyone knew was initiated due to management's lack of trust in staff). Again…no plans. The next question asked how exactly the organization would make the trust and empower value a reality? He said it was all about a mind-set. The author said to him that such a mind-set would not materialize without massive change to decision-rights, employee policies, and so on.

As this clearly was not going to happen, the whole values exercise proved a complete, and expensive, waste of time, because the organization's style of management was based on reinforcing a lack of trust. Without trust, there can be no empowerment. There was one positive outcome of this particularly organization's values program. All employees got a nice coffee cup. This organization's experience is sadly all too common.

Case Illustration: Good Practice – Korea South-East Power

The wholly government-owned Korea South-East Power (KOSEP) is a better example. As part of broader efforts to address poor performance and improve depleted employee morale, the CEO Do-Soo Jang and the executive team introduced four core values to the near-2000-strong workforce: create value, open mind, spirit of challenge, and social contribution. Each of these values was described through both an institution and internal dimension. For example, "spirit of challenge" had institution descriptors around deepening and expanding responsible management and creating a culture of setting goals with the notion of challenge and an internal descriptor focused on motivation of employees.

The CEO has been instrumental in leading and promoting a set of structured activities that have helped the organization internalize a new set of core values. These new values have increased the organization's capacity for change. For instance, to promote the spirit of challenge value, the CEO introduced a 130% Stretch Goal for strategic measures within the Balanced Scorecard as well as an educational program called "Future CEO Training Course" to motivate and challenge employees to develop skills in strategy management.

Advice Panel

When introducing a new strategy, there might be many cultural barriers and resistance to change.

- Fear of measurement/accountability
- Fear of performance transparency
- Notion that this will be another failed initiative so just ignore it until it goes away
- Fear of disrupting the established, and well understood, "rules of the game"

- A legitimate belief that this simply will not work in this type of organization and culture
- A legitimate fear that this will mean more work for already overstretched employees

These fears must be explored and dealt with honestly.

Leadership and Culture

The two contrasting values case illustrations stress the point we made at the start of the previous chapter—leadership and culture are essentially indivisible. Culture is demonstrably led from the top (for better or worse). This is imperative for a culture change effort to succeed. Put simply, leaders must own the culture. Various research studies support this claim.

For instance, one large research program found that the HR department (who will of course have an important facilitative role in rolling out culture change interventions) should not own culture. The research study reported that within organizations with cultures that they themselves described as "well-defined" both culture and values were owned by the senior team in 72% of the cases and by HR in 12% of the cases. For those with "poorly defined" cultures, the respective figures were 40% by the senior team and 50% by HR. [3]

Our field experiences concur that the senior management team (and, most importantly, the CEO or equivalent) must own and champion the strategy-aligned culture and must reinforce the values by starting with their own behaviours. Indeed, nothing will kill a culture change program quicker than leaders espousing one set of behaviours but practicing and rewarding another. Once more, the unwritten rules of the game will be reinforced.

The Shadow of the Leader

A useful technique is to think of the "shadow of the leader" that casts a long way within an organization. From this shadow, employees know (and therefore do) what the leaders really want and reward—not what the leaders might say they want, and even espouse in public statements, internal communications, or even corporate values.

For example, the leaders might encourage individual competitiveness while extolling the virtues of teamwork or encourage going the extra mile for customers while finding ways to punish the same employees for allegedly wasting time and money. If a strategy-aligned culture is to take hold, employees have to believe there will be positive consequences for adopting the new corporate values and behaviours and negative consequences for not doing so. Those employees that most visibly demonstrate the new strategy-aligned behaviours must be publicly acknowledged, celebrated, and rewarded. They must become the new corporate heroes, their actions imbued in corporate stories and folklore. The senior leadership team must make this happen.

Driving Culture Change with the Balanced Scorecard

Although we stress that getting the culture "right" is a critical prerequisite for succeeding with a Balanced Scorecard system, the system itself could be part of a powerful framework for driving cultural change.

Indeed, objectives such as "create a high-performing culture" are frequently found within an organization's learning and growth perspective (see Fig. 11.2), often with supporting KPIs such as living the values and employee engagement. Not useful. Too often, this objective is not defined, nor are other objectives or the KPIs organization-specific. This is a major shortcoming, as "culture" is perhaps the most powerful driver of change on the Strategy Map.

When shaping such an objective, the senior leadership team must take the time to understand what values and behaviours will drive high performance in their organizations, given their corporate strategy. They must understand

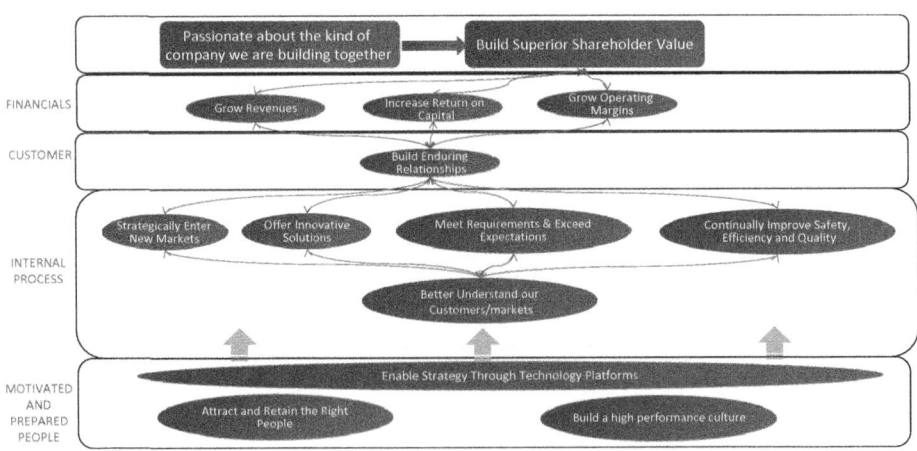

Fig. 11.2 A Strategy Map with a "build a high performance culture" objective

the challenges in moving culture from the current to the desired state and must ensure that appropriate interventions or initiatives are in place to close these gaps. This should begin with ensuring a meaningful objective statement (see Chap. 4: *Strategy Mapping in Disruptive Times*).

When shaping a "culture" objective, take time to think about what type of culture is required. "Develop a Culture of Innovation," for example. Then focus the KPIs of developing this organizational mind-set. Moreover, this can then be part of a strategic theme, thus making the link with internal processes visible. Consider naming the theme something like "Culture of Innovation" to flag that culture is about how work gets done and is not just a people thing. Then learning and growth objectives can be even more specific. A good example of this is the UAE-based automobile retailer A.W Rostamani (see also Chap. 4).

Then consider launching a cultural transformation program to identify and profile the required behaviours of leaders and managers; measure and reward adherence to these values through an effective performance management system; and establish the processes and governance mechanisms to reinforce the desired practices, amongst other goals.

The leadership team must "walk the talk," cast the right shadow, and rewrite and communicate the unwritten rules of the game. A useful step before this is to conduct a cultural assessment.

Cultural Assessment

A cultural assessment can provide early warning signs of cultural barriers to strategy execution.

Senior managers must have a starting point for change, a place where they have undertaken the required analysis to understand the current strengths and weakness of the culture. This will provide a gap analysis from where they are today to where they want to be in the future.

A cultural assessment will also show in which parts of the organization resistance might be fierce and which might be more accepting—and so launch appropriate and targeted initiatives or actions.

To begin the process, we recommend developing a "Cultural Change Agenda," delineating the current and desired states for dimensions such as leadership behaviour, information flows, decision-rights, and so on. For an explanation of change agendas, see Chap. 4.

This should be supported by an assessment of the strategic cultural dimensions (which, on the change agenda, will describe the current state).

A powerful way to do this is through an online cultural assessment tool, which can be arranged according to the identified cultural dimensions and will reveal the areas where improvement is needed enterprise-wide or at departmental levels. Closing this cultural gap should then become a key initiative in fashioning a strategy-aligned culture.

In recent years, a number of useful online cultural assessment tools have been developed. These have evolved significantly since the first generation of the 1990s, which simply enabled answers to questions, to be followed by manual analysis. Today's generation makes use of advanced data analytics to show correlations, predict targeted improvements, and so on, as shown in Panel 2 (see also Chap. 9: *Unleashing the Power of Analytics for Strategic Learning and Adapting.*)

Through such instruments, answers to a series of questions assess the cultural reality against the behaviours, and so on, required to deliver the strategy. If a customer-centric strategy is being implemented, questions might include, "In my department we regularly talk about emerging customer needs," or "we are empowered to make the right decisions for customers without seeking approval." For innovation, questions might be, "In my department we regularly set aside time to brainstorm new ideas," or "we are free to experiment with new solutions without fear of failure."

A cultural assessment instrument can help identify dysfunctional behaviours and processes that are performance bottlenecks deep inside, and across the enterprise. Getting the organizational mind to perform at optimal levels requires attention be paid to both and, it should be stressed, requires leadership to act (reiterating the fact that leadership and culture are indivisible).

Integrating Data

Cultural surveys can provide particularly powerful insights when correlated with other data. A couple of years back, one of the authors was involved in shaping and managing a series of online surveys that, for each of 16 departments, looked at employee engagement, the satisfaction of internal customers, and, where appropriate, the satisfaction of external customers. What was striking about the findings was that the ranking for employee engagement was almost identical for internal customer satisfaction. The top-performer for both, and by some distance, was the same department. The same held true across all three surveys for those that had external customers. Validation, if you will, that the more impactful way to improve the customer experience is to ensure a great employee experience—and this is extremely cultural.

This data triggered specific departmental interventions. We knew where the issues were. Interestingly, one department that had scored near or at the bottom of each ranking on two subsequent surveys improved significantly when the next survey was conducted (indeed, moved into the top half of 16 departments). What was clear from the analysis was that there was a significant leadership issue. Reshaping the management team and replacing with one that was much more empowering and trusting led to a massive change in the performance of the department (and without replacing any of the staff) and the corresponding employee engagement and internal satisfaction scores.

As much as anything, this highlighted that although the senior team is responsible for owning the culture and setting behavioural expectations, it is the front-line managers that turn such aspirations into reality. As the old cliché goes, "Employees do not leave organizations, they leave managers." And the fact is few "leave a CEO," as in any reasonable sized organizations they will have little, if any, exposure to the CEO and the others in the senior leadership team. Do not underestimate the power of front-line managers!

Advice Snippet: **Ten Cultural Commandments for Creating a Strategy-Aligned Culture**

1. Cultural change must not be done in isolation, but rather, hardwired to the strategic goals of the enterprise.
2. Do not set out to create a strong culture, but an adaptive culture: strategies evolve over time and so must cultures.
3. Consider the use of cultural assessment tools to understand the gap between the current and the desired cultures.
4. Formulate values that are meaningful and really do drive organizational and behavioural change.
5. Ensure that policies, procedures, and decision-rights support the espoused corporate values.
6. Culture change is impossible without demonstrable commitment and ownership from leadership and aligned behaviours: they must cast the right shadow.
7. Celebrate cultural heroes who demonstrate the new values and behaviours.
8. Ensure that values are considered in all decision making and public actions.
9. Identify and profile the new behaviours required by leaders and managers and ensure that these are reinforced through incentives, processes, and governance mechanisms.
10. Use the Balanced Scorecard and strategy execution framework to drive the instilling of a strategy-aligned culture.

Parting Words

This and the previous chapter explained the critical role of leadership and culture in enabling strategic success. We explained they are indivisible and very much based on the behaviours of the senior team. When leadership and culture are working in unison and positively, the likely outcome is engaged employees. As we explain in the next chapter, employee engagement is more complex than simply a score on an annual employee satisfaction survey.

> **Panel 1: Putting People First – How Values Drive Strategy Execution at Statoil**
>
> Oil and gas giant Statoil takes a values-based approach to managing the organization. Values are non-negotiable words and actions that drive how the organization performs and behaves, and are central to all management processes.
>
> Values are tightly aligned to the about 700 Balanced Scorecards that Statoil calls Ambition to Action (see also Chap. 6: *Driving Rapid Enterprise Alignment*), which are in place organization-wide. Among other innovations, the Ambition to Action has a "people and organization" perspective (its version of learning and growth) at the top of the scorecard, and devolved scorecards focus on "translation" rather than cascading.
>
> Moreover, inculcating a strategy-aligned and performance-driven culture has also led Statoil, to abandon the annual budgeting process (see Chap. 7: *Aligning the Financial and Operational Drivers of Strategic Success*).
>
> There are many reasons why this is the case. "There are much better ways to manage an organization than through an annual budget," says Bjarte Bogsnes, Statoil's Senior Advisor, Performance Framework. He continues that abandoning the annual budget was critical to demonstrate that Statoil is a values-based organization. "It's meaningless to talk in your values about team-working, if everything in incentives is around individual bonuses against an annual budget," he says.
>
> Statoil has four values: courageous, open, hands-on, and caring—and is committed to incorporating values into everything that it does by ensuring that these values are not just "nice-sounding words" that hang on walls and are essentially ignored. "Kicking out the budget was one way of demonstrating that we were serious about values," says Bogsnes. So how does Statoil manage such a large organization "budget-free," and what role do values play?
>
> Ambition to Action is the central framework for managing the organization, as well as being a framework for implementing strategy.
>
> Ambition to Action is seen as a crucial tool for activating values as well as people and leadership principles, balancing alignment around strategic direction and common business processes with empowerment and local business responsibility.
>
> Interestingly, Bogsnes stresses that they deliberately do not push shareholder value as the primary purpose of the organization, as this is not something that ignites the majority of employees. Something bolder and with a bigger purpose

is required. "Much of this is about using the Ambition to Action process to activate what we are saying about purpose, leadership, values," he says. "All Ambition to Actions in the organization start with an ambition statement, each of which is a translation of the corporate statement to be "a globally competitive and exceptional place to develop and perform.""

People Partnership
The Statoil Book (which can be downloaded from the Statoil corporate website, www.statoil.com, and which provides an overview of how Statoil is managed) stresses the importance of creating a "People Partnership." "[We will] establish and grow a partnership between our group and the individual based on clear expectations and a mutual commitment to the way we behave, deliver, and develop."

Top of what employees should expect from the group is to "promote a stimulating work environment guided by our values and a commitment to your personal and professional development," whereas top of what Statoil expects from employees is to "live our values in all aspects of your work."

Being Courageous
Another group expectation of the individual is to "take the initiative and look continuously for ways to improve performance." This expectation speaks directly to Statoil's value around being "courageous," which has amongst its descriptors to "use foresight, and identify opportunities and challenges," and to "challenge accepted truths and enter unfamiliar territory."

Living up to the "courageous" value is one reason why the idea of being budget-free took hold within Statoil. "Employees appreciate working for an organization that questions established approaches and [is] willing to challenge [them]," says Bogsnes. Moreover, this value has helped drive a culture in which innovation is the norm, as Bogsnes explains: "People are always looking for new ways of doing things, are encouraged to do so and are not afraid to speak up." Indeed, a 2011 ranking by Fortune placed Statoil as the seventh most innovative company in the world. This was the highest in their sector and beaten only by the likely top performers such as Apple, Google, Nike, and Amazon. "Employees are impatient in wanting to improve and are fully empowered to do so."

Management Processes and Culture
Empowerment is, of course, very cultural. Bogsnes stresses the importance of getting the management processes right if a strategy-aligned and performance-driven culture is to be embedded. "If you have management processes that treat employees as potential criminals you will have a culture that reflects this," he says. "There is an important relationship between management processes and the culture you end up with and this is something to be aware of." He continues that, often, what happens is that culture and leadership is seen as an HR responsibility, whereas budgeting is owned by finance. "If these two functions don't align, then the organization ends up with inconsistent messages," he says. "Oftentimes HR will preach theory X leadership while finance is pushing theory Y management processes. At the front line, where these messages are received, employees quickly become cynical and disengaged."

To explain, according to a theory introduced in 1960 by Douglas McGregor at the MIT Sloan School of Management, theory Y managers believe that, given the proper conditions, employees will learn to seek out and accept responsibility and exercise self-control and self-direction in accomplishing objectives to which they are committed. Theory X managers believe that individuals are inherently lazy and not fond of their jobs—very much in keeping with the views of Frederick W. Taylor (see Chap. 1). As a result, an authoritarian management style is required to ensure that individuals fulfil their objectives and do not misbehave. Bogsnes comments, "All companies will have some type X employees, and you must watch out for that. But we firmly believe that most employees conform to the type Y, and we structure the organization and decision rights accordingly."

Bogsnes adds that it is not enough to have a good strategy, but that you need a performance management process to deliver the strategy and with the necessary autonomy, as strategy implementation cannot be micromanaged. "If the strategy is clear and well understood by employees, then they don't need micromanagement as they know the direction and will figure out how to deliver."

Change Management
Statoil also sees change as a normal part of everyday work, and not a separate exercise. Statoil's senior team constantly makes the case for change, creating the necessary discomfort with the current situation. "The better the job we do on this the easier the change," explains Bogsnes. "Change can be an integrated part of how people work and not a project," adding that when an organization needs projects to drive change, it might be an indication that this is not the way of working in the organization.

Results
With such an unusual and pioneering approach to managing the organization, the bottom line is how successful a values-based approach to management has proven to be. "The way we measure financial success is relative to our peers," explains Bogsnes. "We have been doing very well against our peers for many years." For instance, looking at stock performance, in the past decade, Statoil has outperformed most of its European peers and held its own against the giant Exxon Mobil [4].

Moreover, an online survey of more than 10,000 engineering and business students from 29 universities and colleges in Norway in 2014 ranked Statoil the number 1 company to work for—the 18th time it has received this accolade from engineering students and the 13th time from business students [4].

As the final words in the FT.com article, the then soon-to-be departing CEO Helge Lund (who had been in charge for more than a decade) noted about Statoil, "It has everything: good people, politics, geopolitics" [5]. It is likely not accidental that he listed good people first.

Panel 2: Driving Financial Performance Through an Understanding of Culture

Eggi™ is an online assessment tool that, beginning with culture, is geared toward enhancing a workgroup's efficiency and effectiveness while also serving as a development plan for a manager's personal growth. Created by the US-headquartered SurveyTelligence and powered by InfoTool™, it enables demographic segmentation, cross tab analyses, alignment measurements, predictive correlation analyses, and other data assessments [6]. Figure 11.3 shows the overall schematic of the system, with alignment of the organizations being the central goal. Green scores are best practice, yellow mediocre, and red requires immediate attention.

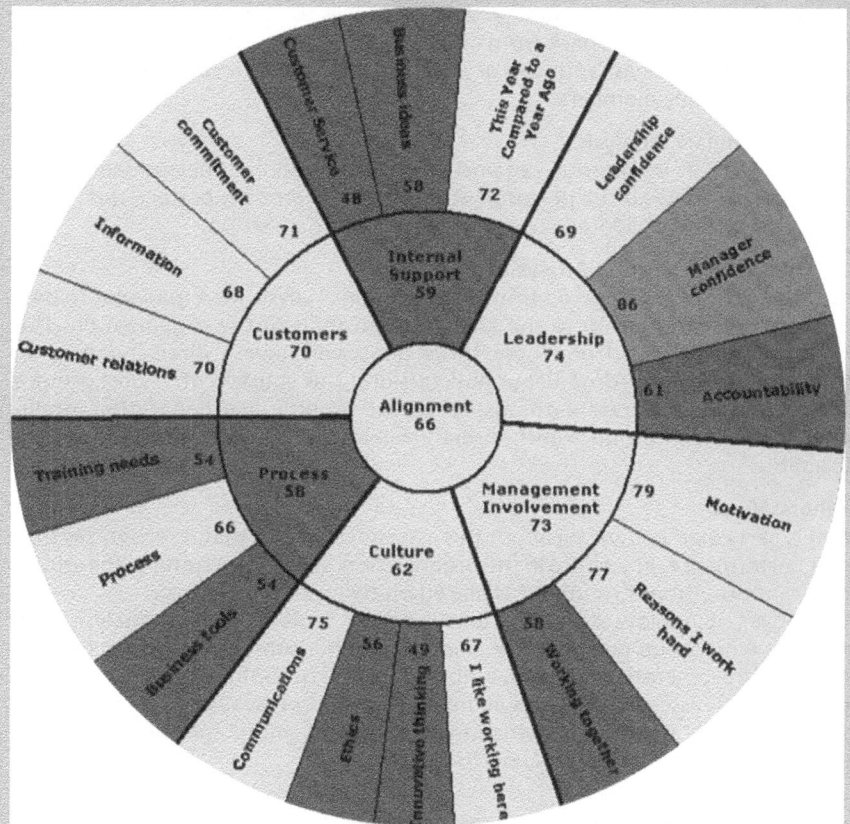

Fig. 11.3 High-level Eggi schematic, showing alignment as the core focus. (Source: SurveyTelligence)

Below the high-level schematic, the online system comprises five report icons, each of which has a different purpose, but work together to drive performance improvement.

Icon 1: Culture Analysis
A culture analysis of a workgroup is generated. This provides a picture and a score of how well the workgroup is performing against agreed business measures. This is complemented by a computer-generated way to take action by simply clicking a "wedge"—the business systems impacting the efficiency and effectiveness of the workgroup.

Icon 2: Continuous Improvement Analysis: From Culture to Action
Computer selected prioritized action steps are chosen to maximize the blueprint for improving of the efficiency and effectiveness from identified workgroup strengths and weaknesses.

Icon 3: Employee Engagement Analysis
From action to people taking the action, responses to all open-ended questions provide unrestricted feedback from a team as to willingness to take action, process improvement ideas, and why or why not they would recommend their job to friends or family members.

Icon 4: Benchmark Analysis
Comparing your workgroup against the organizational performance—this benchmark measures if the performance of the workgroup is better, equal to, or worse than the consolidated scores of the entire organization.

Icon 5: Driver/Correlation Analysis
Three key actions are selected from a correlation analysis for a manager to better align their workgroup to the key strategic goals of the organization. The organizational impact from these three goals will have a direct bearing on achieving the overall organizational mission. This impacts the workgroup's contribution to corporate earnings, corporate growth, and generating intelligence for the workgroup, which directly will impact the organization's strategic goals.

Authors' View
What is particularly powerful about this assessment tool is that it recognizes that great performance, anywhere in the organization, starts from culture. Therefore, significant attention throughout is placed on fixing cultural issues.

Moreover, it deploys advanced data analytic tools to collect, analyse, and report on workgroup performance. Reports, comparisons, along with actions to prioritize, are generated instantly, with a high level of confidence with regard to precision. Such assessment tools will be more and more available and will begin to revolutionize how we understand culture and, more specifically, how it impacts organizational financial performance.

Self-Assessment Checklist

The following self-assessment assists the reader in identifying strengths and opportunities for improvement against the key performance dimension that we consider critical for succeeding with strategy management in the digital age.

For each question, any degree of agreement to the statement closer to one represents a significant opportunity for improvement.

Please tick the number that is the closest to the statement with which you agree		
	7 6 5 4 3 2 1	
In my organization, the attitudes and behaviours the corporate culture promotes are very well aligned to the requirements of strategy execution		In my organization, the attitudes and behaviours the corporate culture promotes are very poorly aligned to the requirements of strategy execution
Our values are identifiably appropriate for our organization		Our values could easily be applied within any organization
There is very good buy-in to the corporate values in my organization		The corporate values in my organization are generally ignored
Organizational structures, decision-rights, and policies generally support the corporate values very well		Organizational structures, decision-rights, and policies generally do not support the corporate values
The senior leadership team demonstrates the values very strongly in their day-to-day behaviours		The senior leadership team demonstrates the values very poorly in their day-to-day behaviours
We have very good tools for assessing corporate culture		We have very poor tools for assessing corporate culture
My organization has a very strong process for closing identified culture gaps		My organization has a very weak process for closing identified culture gaps

References

1. James Creelman, Jade Evans, Caroline Lamaison, Matt Tice, *2014 Global State of Strategy and Leadership Survey Report*, Palladium Group, 2014
2. Rosabeth Moss Kanter, *Change Masters, Innovation and Entrepreneurship in the American* Corporation, Simon & Schuster, 2005
3. James Creelman, *Driving Corporate Culture for Business Success*, Business Intelligence, 1999
4. James Creelman, Jade Evans, Sebastian Rubens y Rojo *Putting People First: How Values Drive Strategy Execution at Statoil*, Palladium white paper, 2015
5. *FT. Com*, 2014
6. See www.surveyintelligence.com

12

Ensuring Employee Sense of Purpose in the Digital Age

Introduction

We are endlessly reminded that the Millennial Generation (exact date-range contested, but for the purpose of this chapter, we use the definition of being born between 1982 and 2000) are changing everything about how organizations work as a consequence of their growing up at the same time as the nascent Internet. The first of these are now in middle management positions and, in some cases higher, especially in businesses that are digitally driven (Fig. 12.1).

The first of the post-Millennial Generation (or Generation Z) will enter university in 2019, and many were born to Millennials. Generation Z are even more comfortable with technology (perhaps knowing how to do a Google search before they could walk!). So, what does this all mean for organizations?

Gallup Research Evidence

A 2016 Gallup report on the Millennial Generation provides some clues. The study reports that 21% of Millennials had changed jobs within the previous year, which is more than three times the number of non-Millennials. Gallup estimates that millennial turnover costs the US economy $30.5 billion annually [1].

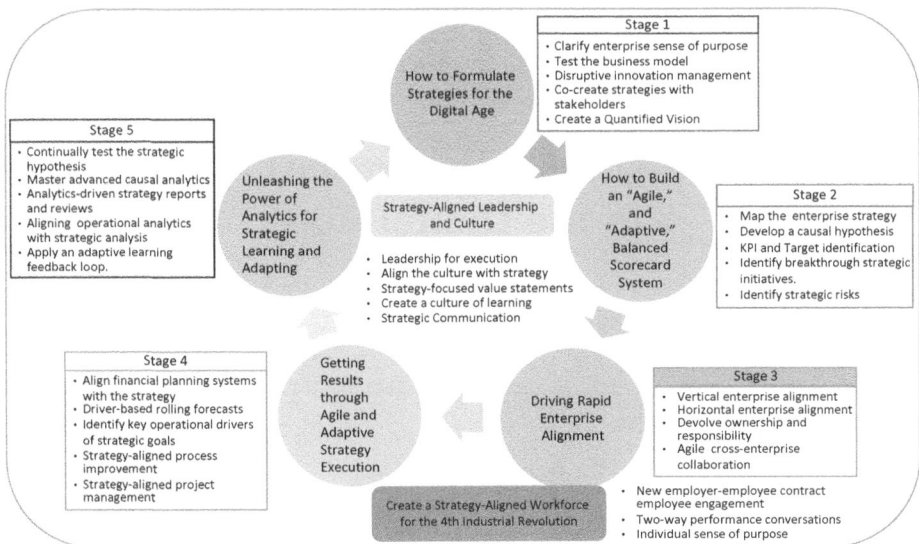

Fig. 12.1 Create a strategt-aligned workforce for the 4th industrial revolution

In searching for clues for the lack of loyalty with this generation (which, as we explain later, is not necessarily something that organizations should always be overly worried about), Gallup also reported that, of the many potential reasons, low engagement in the workplace could be a significant contributor. This is worrying.

Gallup has found that only 29% of Millennials are engaged at work, meaning only about three in 10 are emotionally and behaviourally connected to their job and company. A further 16% of Millennials are actively disengaged, meaning they are more or less out to do damage to their company. The majority of Millennials (55%) are not engaged, leading all other generations in this category of worker engagement.

"The millennial workforce is predominantly" checked out "—not putting energy or passion into their jobs," the report notes. "They are indifferent about work and show up just to put in their hours." Perhaps tellingly, Gallup states that it is possible that many Millennials do not want to switch jobs, but their companies are not giving them compelling reasons to stay. "While Millennials can come across as wanting more and more, the reality is that they just want a job that feels worthwhile – and they will keep looking until they find it." This brings us to an examination of "sense of purpose," which we discuss later in this chapter. But first, we must change the fundamentals of the employee-employer relationship.

Changing the Employee-Employer Relationship

What is required is a fundamental, indeed transformative, shift in the very essence of how we view the employee-employer relationship. As with our knowledge of low engagement scores, this realization has been around for some time.

As far back as 2002, one of the authors of this book wrote a book called *Corporate Culture: Creating a Customer-Focused Financial Services Organization* [2]. For this work, he interviewed Hubert Saint-Onge (now Principal of Saint-Onge Alliance, but then Executive Vice President, Strategic Capabilities, for Canada's Clarica Insurance). A then-pioneer in knowledge management and intellectual capital, Saint-Onge commented, "We must have the mind-set that work is learning and learning is work and that the conventional employee contract is no longer fit-for-purpose.". "We should not expect staff to remain with the organization long-term, but rather create the environment where they learn and so grow their own market-value in exchange for contributing their best in the workplace and therefore to the enterprise." The new employee contract was about capability building and application.

More recently, in his 2016 book *Superbosses: How Exceptional Leaders Master the Flow of Talent*, Sydney Finkelstein observed that, "The best bosses are innovators in how they think about talent. That's true for how they motivate, inspire, and coach their team members, but it also goes for how they think about employee churn. Not only do great bosses not fear people coming and going in their organizations, they actually embrace it" [3].

Finkelstein continues, "How many people entering the workforce today are looking forward to getting that proverbial gold watch after 25 years? Instead, companies are being confronted with a generational mind-set shift that values learning and engagement much more than job security. In this world, trying to optimize around talent retention—as many companies do—can hurt more than it helps."

This means that organizations should think very differently about managing "talent." Indeed, the very word "talent" is somewhat disturbing. It suggests that a small cadre of employees are talented, and the rest are not—hardly motivational.

Moreover, employee engagement scores might be stubbornly low, partly because employees often just reach the end of the road regarding the value they get from the organization and so begin to disengage. From our observations, we would say that when an individual feels they have plateaued and might benefit from another job, this should be managed proactively by managers and HR and thus avoid employees surreptitiously looking for new jobs, disengaging and then, when alternative employment is secured, resigning. Paradoxically, in certain cases proactively managing the "leaving" process might lead to the retention of talent, through redeployment, and so on.

Looking forward, we need to show bravery and jettison notions of employee loyalty or worry incessantly about high levels of retention (unless employee disengagement is the cause) (a mind-set change for HR, which is often judged on the length of time a recruit stays).

We should create an internal workplace dynamic where processes are in place to help the individual leave the organization to further their own development if they cannot gain these competencies or experiences elsewhere in the organization. Some companies encourage an employee to leave and then welcome them back several years later, with an improved toolkit of skills and new experiences at their disposal. Keep in mind that the goal should be that when an employee leaves the organization, they still feel a sense of loyalty to the organization and the brand. These days, it is not uncommon for candidates to approach ex-employees (through LinkedIn, etc.) to gain a reference for the company to which they are applying. A great recommendation from an ex-employee is a powerful recruiting tool.

A study from 2013 had a telling finding. Whereas Google had an average employee age of 29, average tenure was 1.1 year (one of the lowest in the study). Kodak topped the list with an average age of 50 with a tenure of 20 years [4]. One must wonder whether Kodak's HR people got a large bonus for success in employee retention and Google's team was fired (despite being the 6th best company to work for on the planet that year, according to one study) [5].

The End of Appraisals

As well as the misplaced obsession with "employee loyalty," another love of many HR functions is the annual employee appraisal. Along with the annual budget (see Chap. 7: *Aligning Financial and Operational Drivers of Strategic Success*), this is the most ridiculous ritualized dance conducted by organizations. If we asked ourselves the simple question, "What must we do to get great performance from our people?" The answer would not be the budget or an appraisal.

As Human Capital author Liz Ryan rightly said in the Forbes article, *Five Outdated Leadership Ideas that need to die*, the annual appraisal is more about reward and punishment than employee motivation, [6].

"Many managers were taught that people are motivated by rewards and punishments, but that's ridiculous. Maybe donkeys can be motivated with carrots and sticks, but not the brilliant humans on your team. You don't have to do a thing to motivate a human being, except create a workplace in which people feel safe bringing themselves to work."

She continued, "You have to take away the stupid rules and constant measurements if you want to see greatness from your team. You have to get rid of all grading systems that give grown adults A, B and C grades like little kids in school." Indeed, one of the authors of this book once saw a highly engaged employee with several Masters' degrees leave an appraisal in tears. Unsurprisingly, she disengaged and left the company soon afterwards.

The Views of Dr. Deming

However, note that the inappropriateness of the appraisal system is not a recent, millennial-inspired phenomenon. Total Quality Guru Dr. W. Edwards Deming was a fierce opponent many decades ago. In his book *Out of Crisis* he said, "The idea of a merit rating is alluring. The sound of the words captivates the imagination: pay for what you get; get what you pay for; motivate people to do their best, for their own good. The effect is exactly the opposite of what the words promise" [7].

He argued that the appraisal system encouraged short-term performance at the expense of long-term planning, discouraged teamwork, and created a system whereby to get a promotion (or a pay rise) you need a short-term hit. He concluded that the appraisal system led to employees living in fear and being in constant competition with each other.

Deming elaborated, "The fact is that the system that people work in and the interaction with people may account for 90 or 95 percent of performance." Couple this with Deming's declaration that between 85 and 97% of problems in an organization are the responsibility of management, we get to see a fundamental failure of appraisals—it's the wrong way around, we could argue!

The Dangers of Assigning KPIs to Individuals

The views of Deming point to the shortcomings of assigning KPIs to individuals (the manager's expectation of the reporting employee's performance over the next period and that will likely appear on a "personal scorecard," and significantly impact the annual bonus).

Now there are huge issues here. First, few managers (or organizations, for that matter) have any idea how measures work and the variables that impact them (see Chap. 5: *How to Build an Agile and Adaptive Balanced Scorecard*).

The fact is that in a complex system such as an organization, no individual (or team) can have a KPI that is not significantly affected by the behaviour and performance of others. And unless there's close cooperation between managers across departments, one employee's/team's KPI might only be achievable at the expense of somebody else's. Not clever!

Moreover, being able to evaluate gradings from different managers with different personal and cultural biases in a meaningful way is simply impossible. Deming recognized that there are too many variables in human behaviour to make performance comparisons statistically meaningful. Keep in mind that Deming was, first and foremost, a statistician.

Changing the Conversation

We need to move on from performance evaluation to performance discussions based on a structured annual or bi-annual system.

Here's a typical "discussion." A nervous employee walks into his room where his/her supervisor/manager is, who essentially tells them whether they have been a good boy or girl over the previous period, or a bad boy or girl. As such events are these days cleverly disguised by HR as a "constructive two-way conversation," the "child" is asked for feedback from their all-knowing mother or father. But how many reply with, "Yes I have performed poorly this year by my standards because you are such a useless boss." Precisely!

We need to replace this with an ongoing dialogue about the goals of the individual, team, and organization and how these align. In the digital age, much of this can be online, although regular, often ad-hoc discussions should take place.

As well as seeking input from other team members and colleagues elsewhere, this performance dialogue should be as much about the leader as the direct report. Of course, this requires a leader that can live by these norms, which is rare in our experience.

Indeed, The Palladium Group's 2014 Global State of Leadership Report found that of almost 1300 respondents, just one in three believed that their leaders prefer employees who challenge them [8]. This means we need to rethink what we think of as great leadership. A great leader encourages their people to tell them what they think, what they as leaders are doing wrong.

This takes us back to behaving like equal adults and not basing it in a parent/child relationship. It is also about trust.

Theory X and Theory Y

Trust is one of the favourite words used in value statements (see the previous chapter). In the 1960s, MIT professor Douglas McGregor explained theory X and theory Y management styles. Theory X managers believe that employees will do all they can to avoid "work" and must therefore be distrusted and controlled (much in keeping with the thinking of Fredrick W. Taylor). The appraisal system reinforces this.

Theory Y managers believe that, given the proper conditions, employees will learn to seek out and accept responsibility and exercise self-control and self-direction in accomplishing objectives to which they are committed. Clearly still not the norm in most organizations. The Palladium research found that just 32% of survey respondents believe that employees are granted autonomy over how they should fulfil their goals in their organizations. "This startling lack of trust that leaders display has a significant impact on morale and performance," says Jade Evans, Consulting Manager at Palladium. "Our research shows those organizations that do value 'unconventional' approaches in their teams are 3.9 times more likely to outperform their competitors in terms of organizational performance." This is as damning as saying that "trust" is still considered "unconventional."

A Sense of Purpose

Deming talked about intrinsic motivation, getting people to do their best because they see the personal value of their work and, from that, their contribution to their team's and organization's goals. We can update this today to think of sense of purpose—align the individual's sense of purpose with that of the team and the organization.

Organizational Sense of Purpose

As we explained in Chap. 2: *From Industrial- to Digital-Age Based Strategies*, an organization's sense of purpose is captured in the mission statement, such as

Google's, "To organize the world's information and make it universally accessible and useful." The mission describes why the organization exists.

Individual Sense of Purpose

Millennials are clearly looking for a very different workplace experience than earlier generations. For the individual (and this is not limited to Millennials), sense of purpose has two dimensions.

The first is about what they want to achieve professionally and while with the present organization. Worryingly, Palladium found that the vast majority of survey respondents showed a palpable lack of connection to or even interest in their work. In fact, a startling 81% were not content in their role, nor did they feel that their work enables them to fulfil their aspirations and ambitions. The number of respondents who felt this dropped by a massive 33% in organizations that are outperforming their competitors.

The second dimension, which is increasingly evident in Millennials and Generation Z, is the desire to work for an organization that shares their commitment to being better citizens (just as these groups were raised in the internet age, this also coincided with a growing movement focused on environmental and social responsibility).

Deloitte Research Findings

In some quarters, the focus on environmental and social responsibility has led to Millennials being labelled as "anti-business." The Deloitte 2017 Millennial Survey (which is based on the views of about 8000 Millennials across 30 countries) shows this to be a myth, with 76% of Millennials (and 89% of those characterized as "super-connected," the highest users of social media, etc.) being pro-business and believing businesses make a positive impact on the wider society. Furthermore, 62%, consider business leaders as committed to helping improve society (a nine-point increase since 2015) [9]. We consider how Balanced Scorecard systems are being used to drive "positive impact" in the next chapter.

A key message of this book is that organizational structures and working practices must change for the digital age. This certainly resonates with Millennials. As the Deloitte report says, "It is in the workplace where Millennials feel most influential and, in turn, accountable. This is an important point for businesses to acknowledge as it offers a platform from which to build each employee's sense of purpose and, ultimately, a more engaged workforce."

Millennials, the report stresses, believe they have the greatest level of accountability for, and influence on, client satisfaction. Across the "most important" aspects that were measured, perceived levels of accountability was very closely correlated with influence. Other aspects included working culture/atmosphere, general processes/ways of working, and, interestingly, overall reputation of the company.

These finding echoed Deloitte's 2016 millennium survey that suggested that organizations taking an inclusive approach, rather than an authoritarian/rules-based approach, were less likely to lose people. It is also found that employee satisfaction was high in 76% of organizations taking a "liberal/relaxed" approach to management against only in 49% of the more controlling, rules-based organizations.

There are other findings in the 2017 survey that suggested that Millennials prefer working in a collaborative and consensual environment rather than one that directly links accountability and responsibility to seniority (or pay). Although two-thirds (64%) would like their senior leadership to take on higher levels of accountability (as we stated earlier, hierarchies still matter), the majority also believes that people should either take collective responsibility (16%) or—irrespective of their positions or salaries—as much personal responsibility as possible (47%).

As the Deloitte report's authors noted, "Such flexibility is regarded by Millennials as having a positive influence on each aspect of work we enquired about. Tellingly, they say that flexible working arrangements support greater productivity and employee engagement while enhancing their personal well-being, health, and happiness. Compared to those in 'low-flexibility' environments, those employed where flexible working is highly embedded are twice as likely to say it has a positive impact on organizational performance and personal well-being."

Moreover, accountability and flexibility are highly correlated. Those working in the more flexible environments report higher levels of personal responsibility. For example, where flexible working is most deeply entrenched, 34% take "a great deal" of personal accountability for their organizations' reputations. This compares to just 12% within enterprises where there is low flexibility.

In direct opposition to the established thinking that underpinned the *Scientific Principles of Management*, the report states that any misgivings that the opportunity would be abused or that productivity might suffer appear to have been unfounded. "There is clearly potential for employees to feel colleagues are taking advantage of flexible working opportunities, or for line managers to be suspicious of those who regularly work from home or vary the start and finish of their working day."

"However, the potential for a distrustful atmosphere is largely unrealized with three-quarters (73 percent) of those offered flexible working opportunities saying they trust colleagues to respect it. An even higher proportion (78 percent) feel trusted by their line managers. Perhaps as one would expect, where flexible working is most embedded, the levels of trust are greatest with only one in 10 suspicious of colleagues or believing that their line managers doubt them."

The key words from the Deloitte survey are that Millennials are seeking collaboration, accountability, influence, flexibility, and trust. This has significant implications for organizations and those that still prefer Theory X-type management styles.

Team Sense of Purpose

As a final element of "sense of purpose," this also exists at the team level.

Here, we are considering how the team interacts with the organization, other teams, and the individual, and how the enterprise engages team members in the execution of the strategy.

An overwhelming majority of organizations studied by Palladium were not satisfied with the level of cooperation among their teams. Eighty-seven per cent reported that people within their organization fail to cooperate, support, and care for one another. Moreover, almost three out of four experience feelings of anxiety or distress that do not subside rapidly after work problems or disputes are resolved. This is alarming in many ways, but particularly in light of the fact that, according to Palladium's research, cooperative teams make organizations 3.7 times more likely to be among the top performers of their industry. It also confirms Gallup's finding that actively disengaged employees experience high levels of anger and anxiety while completing their daily work. Anger and anxiety are the staple diet of the invisible elephant.

Communication

Along with the organizational, team, and individual dynamic, a further area that requires attention is communication. The old cliché, "communicate, communicate, communicate" is more important than ever—and more challenging.

From a strategy viewpoint, research has found that here is a correlation between organizational performance and how well the strategy is communicated to employees. According to one study, 67% of staff in organizations that were "well performing" had a good understanding of the overall organizational goals, compared to 38% in those that were "poorly performing."

Effective strategy communication program has four components.

1. Define Target Audiences
Strategy has to be communicated to different audiences, with different information requirements, be they external (shareholders, regulators, customers or suppliers, and even communities, for many organizations) or internal (board of directors, executive leadership, middle management, frontline staff).

2. Identify the Message Streams
A Message Stream is a continuing flow of information on a given subject, with the goal of conveying to internal and external audiences how various efforts can impact the strategy. There are four types of message streams.

(a) **Strategy**: inform audience about strategy-related activities
(b) **Success**: define what represents success in the coming year
(c) **Results**: report results from strategy execution that communicate progress
(d) **Impact**: explain how each employee's job impacts strategy.

Although different wording and information are provided to different audiences, it is important that they tell "the same story." In today's fully connected world, any differences in the core messaged will "leak" and will stymie implementation efforts and potentially damage the reputation of the organization.

3. Select and Design Communication Channels
There are many communication channels available today, from town-hall style meetings, individual face-to-face meetings through newsletters to emails and social media. Again, ensure the messages are consistent.

4. Measure, Solicit Feedback, and Foster Learning
Thanks to digital connectivity, gone are the days when organizations communicated internally and externally on a one-way street. Communication must evolve from a "PR exercise" to a way to begin performance conversations and solicit feedback. Millennials, in particular, expect a two-way and even group dialogues, as well as differences in opinion. There are myriad ways to do this, such as surveys for each of the target audiences to online employee and customer chatrooms. What is important is that feedback is acted upon, with the audience informed as to the outcome of their suggestions.

As shown in Panel 2, communication can also be guided by the 5 Cs of Clarity, Credibility, Concision, Context, and Consistency.

Essex Police Cast Illustration

The 6000-employee-strong, UK-based Essex Police recognized the importance of a robust communications strategy when it built a Strategy Map (which it called a plan on a page) in 2014. In an interview with one of the authors, Mark Gilmartin, Director of the Essex Police and Kent Police Support Service, noted that, from the outset, the senior team of Essex Police realized that if they were to successfully secure a buy-in to this new approach to performance management and measurement, it would require a comprehensive communication strategy. "As much as anything this was the recognition that adopting the Plan on A Page necessitated an evolution of its performance culture," he says.

Early on, an external facilitator conducted workshops with senior managers explaining the principles of best practice on performance management. The communication strategy also included roadshows for operational staff, who included frontline officers from each discipline: neighbourhood, public order, and firearm officers, and so on. This provided them with an opportunity to contribute to the strategic performance conversations and even ask direct questions to the Chief Constable, who personally led the communication effort. "We perhaps over-communicated, but for good reasons as in the Police world front-line people are not used to being fully consulted on strategic matters," says Gilmartin.

One of these "good reasons" speaks directly to cultural challenges. Gilmartin explains that within most UK police forces (also true in many other countries) there is a not insubstantial amount of cynicism, especially when it comes to anything to do with performance management and measurement.

For many Police officers and staff there's a response of, "been here before, done it and nothing changed," he says, stressing that the continued push from the top and continued communications has helped reduce that resistance significantly.

Gillman also notes that the Strategy Map has enabled better performance reporting and discussion. "[The Strategy Map] has enabled a much-improved format for presenting the information in a way that encourages broader discussion about performance – informed by, but not restricted to, the data."

He states that there is now greater usage of statistical evaluation and longer-term trends to provide a more measured view of performance. "As we become more mature in using the performance management framework we will get better at analysing the data that we collect and in how we present those insights to senior management and other teams so that they have much richer and data-driven performance discussions," he says. "And in the final analysis it is the quality of the conversations had and decisions made that counts, not the framework itself" [10].

Human Capital Development to Execute Strategy

In recognition of the importance of culture and employee engagement, Dr. David Norton, has developed a "Human Capital Development to Execute Strategy" framework. This framework, is based on the pillars of engagement (employees understanding the strategy), alignment (personal goals linked to the strategy), readiness (employees have the necessary skills and competencies), and culture (the organization adopts the set of values required by the strategy). This framework plots the causal relationship between these four human capital pillars and ultimate financial or mission success (Fig. 12.2).

The framework emphasizes the importance of leadership development and of employees understanding the strategy. Organizations with strong performance-based cultures have leaders who understand how to execute strategy. They ensure that their employees understand the vision for the

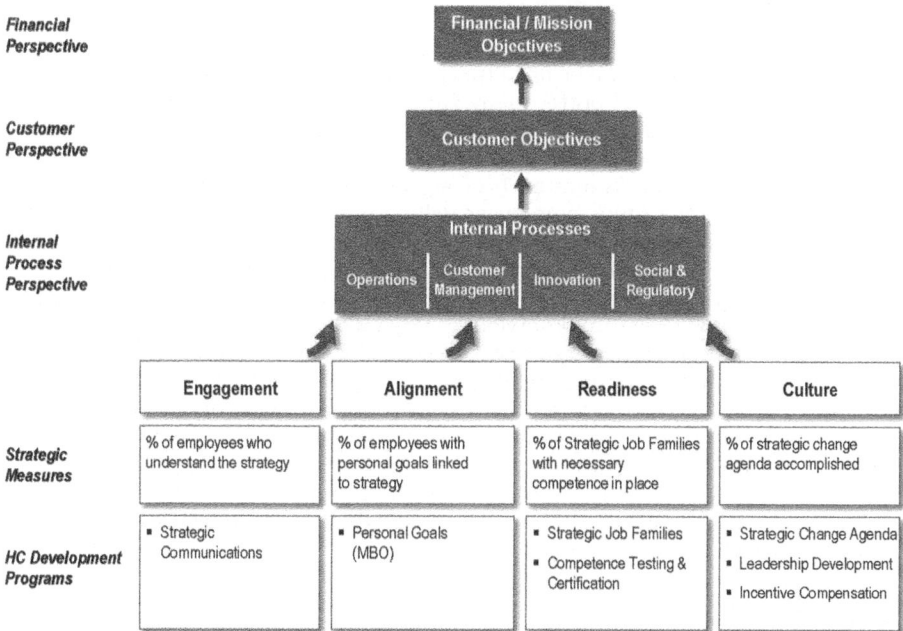

Fig. 12.2 The causal relationship between the four human capital pillars and ultimate financial or mission success

organization and their personal role in delivering that vision and, therefore, the strategy's success.

Parting Words

Those organizations that will create sustainable success in the early decades of the digital age will be those that take a very different approach to the workplace experience than has hitherto been the norm. They will stop obsessing about employee loyalty or retention and rather than view employees as people that "work for them" (and therefore somehow "owned," and so to be emotionally abused in appraisals) view the relationship as "working with them," and perhaps for short times. This is a very different mind-set.

Work will become increasingly virtualized and so the end of the "clock on, clock off" mentality. People have different working rhythms, which will increasingly be accounted for.

These organizations will have, in the words of change author Dr. James Belasco, "taught the Elephant to dance" [11].

Panel 1: Why Do We Pay Bonuses?

One of the key functions of the appraisal system is to assess and provide bonus payments (often based on specific KPIs). First, doing so will likely lead to all sorts of dysfunctional (or, as we prefer to say, rational) behaviour to hit those KPIs. As the old saying goes, "be careful what you ask for, you might just get it." See the story about Mumbai and rats in Chap. 5.

OK, money and benefits matter. They are, as Maslow taught us, hygiene factors [12]. If the pay is lousy, employees will be unlikely to engage. An obvious fact that many organizations still fail to recognize—oftentimes, in pursuit of heightening profits to deliver shareholder value. But pay has lesser importance once the hygiene factors are dealt with, which today is mostly the case, at least in developed markets and for knowledge-based workers.

With some exceptions (typically sales related), stop paying individual bonuses—how does that encourage teamwork, for instance? It might be worth starting with the question, "why do we pay bonuses in the first place?" Many might argue that it is to reward outstanding performance. However, if the performance of one individual is based on the performance of others (who may well be in a different department), then the individual's contribution is very difficult to isolate. Moreover, the performance might be heavily influenced by economic factors, over which the individual has zero influence.

An additional question should be "do bonuses actually work in improving performance?" A research study by the Netherlands-based research firm HPO Center found that "a fair reward and incentive structure did not show a significant relation with organizational performance. The conclusion therefore was that using bonuses or implementing certain types of reward systems does not have a positive or negative effect on long-term organizational performance.

"A possible explanation for this result is that bonuses and reward systems are a hygiene factor for an organization," said report author Dr. André de Waal, HPO Center Academic Director. "If the organization does not have an appropriate reward system, with or without bonuses, it will run into trouble with its employees. If it does, which employees expect and consider to be normal, it can start working on improving its performance".

De Waal concluded that, "Putting a lot of effort in introducing bonuses or a certain type of bonuses and reward systems and then expecting your organization to improve its results and maybe become an HPO [high-performance organization] is unrealistic. The bonuses and reward system is not a determining factor for long-term high performance."

We also need to move away from believing that levels of pay are mostly about hierarchical levels. No. A great technical engineer might also be a great coach and mentor, but might not want to be a manager. Why punish this through the conventional pay = status process?

Panel 2: The Five Cs of Communication

An effective communication program can be based on the following five Cs

1. Clarity
2. Credibility
3. Concision
4. Context
5. Consistency

Clarity
From a strategy point of view, clarity is important because the messages can easily become confused.

When communicating, we must balance the challenges of delivering a message designed to create focus around the goals and objectives of the organization, while simultaneously encouraging all staff to challenge assumptions made during the definition/selection of said objectives. It is critical to keep an "open mind" and continuously consider the risks, both threats and opportunities, around those objectives.

There is also the challenge of setting out a positive and inspiring vision, with a clear set of motivating objectives that are clearly understood (hence the importance of shaping good objective statements, see Chap. 4, *Strategy Mapping in Disruptive Times*).

Credibility
To be effective, not only do the messages need to be credible, but the people and management teams delivering those messages must be equally credible. As

explained in Chap. 10: *How to Ensure a Strategy-Aligned Leadership*, thinking about the "shadow of the leader" creates the culture that determines expected working practices and attitudes. Simply put, subordinates will behave in ways that mirror their leaders.

Concision
It is important that each message regarding strategy and strategy execution should either provide information or explain an action. The purpose of the message should be clear, and it should be easy for those receiving the message to take appropriate action.

Context
All communication must consider the current organizational context and environment. One of the quickest ways that credibility is destroyed, and support lost for an approach, is if messages reaching staff appear to conflict with their understanding of the situation—thus, the messages being communicated come across as being out of context and are unlikely to be believed. This is particularly problematic when the organization is facing economic challenges, such as job losses.

Consistency
Mixed messages and inconsistency will undermine efforts to implement a strategy. Organizations must be consistent in their messages around strategy to ensure staff and other stakeholders are engaged and on-board with the implementation efforts. While changes over time will be necessary, they should be communicated clearly and managed so that support is not lost and people do not become disillusioned about their role in the achievement of company's objectives. Basically, it's about making sure everybody is on the same journey and are not at a different bus stop.

Self-Assessment Checklist

The following self-assessment assists the reader in identifying strengths and opportunities for improvement against the key performance dimension that we consider critical for succeeding with strategy management in the digital age.

For each question, any degree of agreement to the statement closer to one represents a significant opportunity for improvement (Table 12.1).

Table 12.1 Self-assessment checklist

Please tick the number that is the closest to the statement with which you agree		
	7 6 5 4 3 2 1	
My organization has a very good understanding of the workplace expectations of different demographics		My organization has a very poor understanding of the workplace expectations of different demographics
Employees are highly engaged in my organization		Employees are poorly engaged in my organization
My organization is primarily focused on hiring talent for the timeframe that is appropriate to the organization and the individual		My organization is primarily focused on long-term employee retention/loyalty
In my organization, the appraisal system is very motivational		In my organization, the appraisal system is very demotivating
In my organization managers encourage reports to challenge them		In my organization managers discourage reports from challenging them
My organization is more Theory Y (trusting) than Theory X (distrusting)		My organization is more Theory X (distrusting) than Theory Y (trusting)
In my organization, we place great significance on aligning the sense of purpose of the individual with that of the organization		In my organization, we place no significance on aligning the sense of purpose of the individual with that of the organization
My organization has a very good process for strategic communications		My organization has a very poor process for strategic communications

References

1. Amy Adkins, *Millennials: The Job-Hopping Generation*, Gallup, 2016
2. James Creelman, *Corporate Culture: Creating a Customer-Focused Financial Services Organization*, Lafferty Publications, 2002
3. Sydney Finkelstein: *Superbosses: How Exceptional Leaders Master the Flow of Talent*, Penguin Random House, 2016
4. Vivian Giang *Ranking America's Biggest Companies By Turnover Rate*, Business Insider, 2013
5. *The 50 Best Companies To Work For In 2013*, Business Insider.
6. Liz Ryan, *Five Outdated Leadership Ideas*, Forbes, June 2016
7. W. Edwards Deming, *Out of Crisis*, MIT Press, 1982

8. James Creelman, Jade Evans, Caroline Lamaison, Matt Tice, *2014 Global State of Strategy and Leadership Survey Report*, Palladium Group, 2014
9. The Deloitte 2017 Millennial Survey, Deloitte.
10. Bernard Marr, James Creelman: *Implementing a Performance Management Framework at Essex Police Support Service,* Advanced Performance Institute, 2015.
11. James Belasco: *Teaching the Elephant to Dance, Empowering Change in Your Organization,* Penguin Random House, 1990.
12. See https://en.wikipedia.org/wiki/Two-factor_theory

13

Further Developments: Driving Sustainable Value Through Collaborative Strategy Maps and Scorecards

Introduction

In the previous chapter, we explained that the millennial and post-Millennial Generations are imbued with a sense of demanding that organizations (and governments) be more socially and environmentally aware. Increasingly, we are witnessing how an organization's performance as corporate citizens is impacting their reputation and from that consumers' purchasing behaviour. Of course, through social media any perceived breach of Corporate Social Responsibility (CSR) can be broadcast globally in seconds. There's no hiding place, and the ensuing damage to reputation (and from that earnings) can be considerable.

However, the possession of a strong sense of social and environmental responsibility is not the creation of Millennials, neither is it particularly new to how many organizations do business. Many firms were notably charitable in the nineteenth, and early decades of the 20th, centuries (although often viewed negatively, as it was seen as giving away shareholder funds without their consent), but we can see the idea of CSR emerging more formally post-World War II.

Corporate Social Responsibility

As early as 1946, *Fortune* Magazine polled business executives asking them about their social responsibilities. As academic Archie Carroll explained in *A History of Corporate Social Responsibility: Concepts and Practices*, "The results of this survey suggest what was developing in the minds of business people in the 1940s. One question asked the businessmen whether they were responsible for the consequences of their actions in a sphere wider than that covered by their profit and loss statements. Of those polled, 93.5% said yes. Second, they were asked 'about what proportion of the businessmen you know would you rate as having a social consciousness of this sort?' The most frequent responses were in the categories of 'about a half' and 'about three quarters.' These results seem to support the idea that the concept of trusteeship or stewardship was a growing phenomenon among business leaders." [1]

The 1950s saw the wider acceptance of such ideas, driven somewhat by the 1953 publication of Howard Bowen's landmark book *Social Responsibilities of the Businessman* [2]. This was the first comprehensive discussion of business ethics and social responsibility and created a foundation by which business executives and academics could consider the subjects as part of strategic planning and managerial decision making. But clearly, as the title suggests, as does the questions in the previous example, the idea of businesswomen had yet to take root!

Triple Bottom Line

The next major development came in 1994, when John Elkington coined the term, "Triple Bottom Line"—a concept that encourages the assessment of overall business performance based on three important areas: profit, people, and planet. Importantly, key to the model was that each of these areas would be rigorously measured.

The Triple Bottom Line idea emerged around the same time as the Balanced Scorecard, and with both aiming to more formally "balance" financials and non-financials, we witnessed early examples of the models merging and, as a result, starting to position CSR as more directly impacting strategy execution. As with many of the "non-traditional" approaches to management (such as abandoning the budget: see the Statoil example in Chap. 7: *Aligning the Financial and Operational Drivers of Strategic Success*) Scandinavia provided the pioneers.

Nova Nordisk Case Illustration

Headquartered in Bagsværd, Denmark, multinational pharmaceutical company Nova Nordisk introduced a Balanced Scorecard in 1995 that was based on the principles of the Triple Bottom Line. In a 2001 report written by one of the authors of this book, Peter Moeller, then Novo Nordisk's Vice President, Organizational Development, explained that utilizing the Balanced Scorecard framework provided a vehicle for focusing on the critical few performance objectives and targets that would make a difference. "It is an excellent way to systemize the company's tradition of focusing on much more than financial measures," he said [3].

Within the scorecard, the learning and growth perspective included an objective on "equal opportunity." The company decided to focus on this as a critical objective because of its obligations to social responsibility and, over the following three years, worked to ensure that equal opportunities were fully ingrained into the policies and practices of the company. Therefore, equal opportunity objectives, KPIs, and targets were captured on the corporate scorecard and mandated onto scorecards throughout the company. Note that this particular area was chosen partly as a result of the organization recognizing that, although environmental performance standards and measures were well embedded into the company, those for social responsibility were much less so.

More than 20 years later, this integrated model is still in place within the organization: indeed, The Triple Bottom Line principle is now Novo Nordisk's business approach and was included in the Articles of Association in 2004 [4].

In the 2016 Annual Report, Lars Fruergaard Jørgensenn, Nova Nordisk's President and Chief Executive Officer (CEO) said this in his CEO letter, which neatly summarized the organizations sense of purpose and how this aligns with that of the individual and the role that Triple Bottom Line plays in this [5].

"It is not just what we do, but also how we do it that makes Novo Nordisk a special company. The 'Novo Nordisk Way,' describes who we are, where we want to go and the values that characterize our company."

He added that, "Over the years, it has become clear to me that the Novo Nordisk Way is the reason why many of our employees are working here and not somewhere else. It is about always having patients' interests in mind, about always doing what is best in the long run and about doing business in accordance with the Triple Bottom Line business principle, which means that we always consider the financial, environmental and social impacts of our decisions."

Sustainability Strategy Map

Over the past couple of decades, there have been numerous other examples of organizations utilizing the Balanced Scorecard system to drive CSR and developing frameworks to do this. One useful recent spin is the Sustainability

Strategy Map, developed by Canada-headquartered consulting firm Pm2Consulting (Performance Measurement & Management). In a May 2017 LinkedIn article, founder and CEO Brett Knowles, stated that, "We now need to integrate this new best-practice around measuring environmental and societal business impacts with the third 'bottom-line'… business success."

Best practices on measuring business success comes, he writes, from the Balanced Scorecard body of knowledge. In the approach pioneered by Pm2Consulting, this is aligned to the 21 Sustainability Business Goals with specific measures and formulas that are found in the Future-Fit Business Benchmark. The goals, as examples Energy is from renewable sources, Products do not harm people or the environment, Employees are paid at least a living wage and Business is conducted ethically.

The Sustainability Strategy Map uses the same four perspectives as the Balanced Scorecard (for the company goals) and adds in the two additional "perspectives"—society and environment. They exist as columns down the right and left side as the company interacts with them across all four balanced scorecard perspectives. Fig. 13.1 shows a generic example of a Sustainability Strategy Map.

Figure 13.2 shows how the 21 benchmark goals align with the strategic objectives. Knowles comments that, "If we combine these two views – the Sustainability Strategy Map and the Future-Fit Business Benchmark measures, we can see how we can populate each of these goals with hard measures." [6].

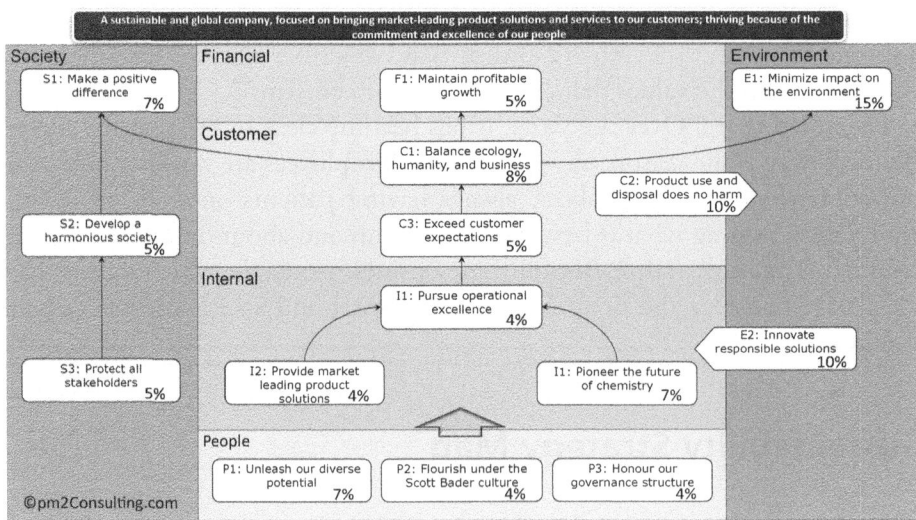

Fig. 13.1 A Sustainability Strategy Map. (Source: pm2Consulting)

Further Developments: Driving Sustainable Value... 247

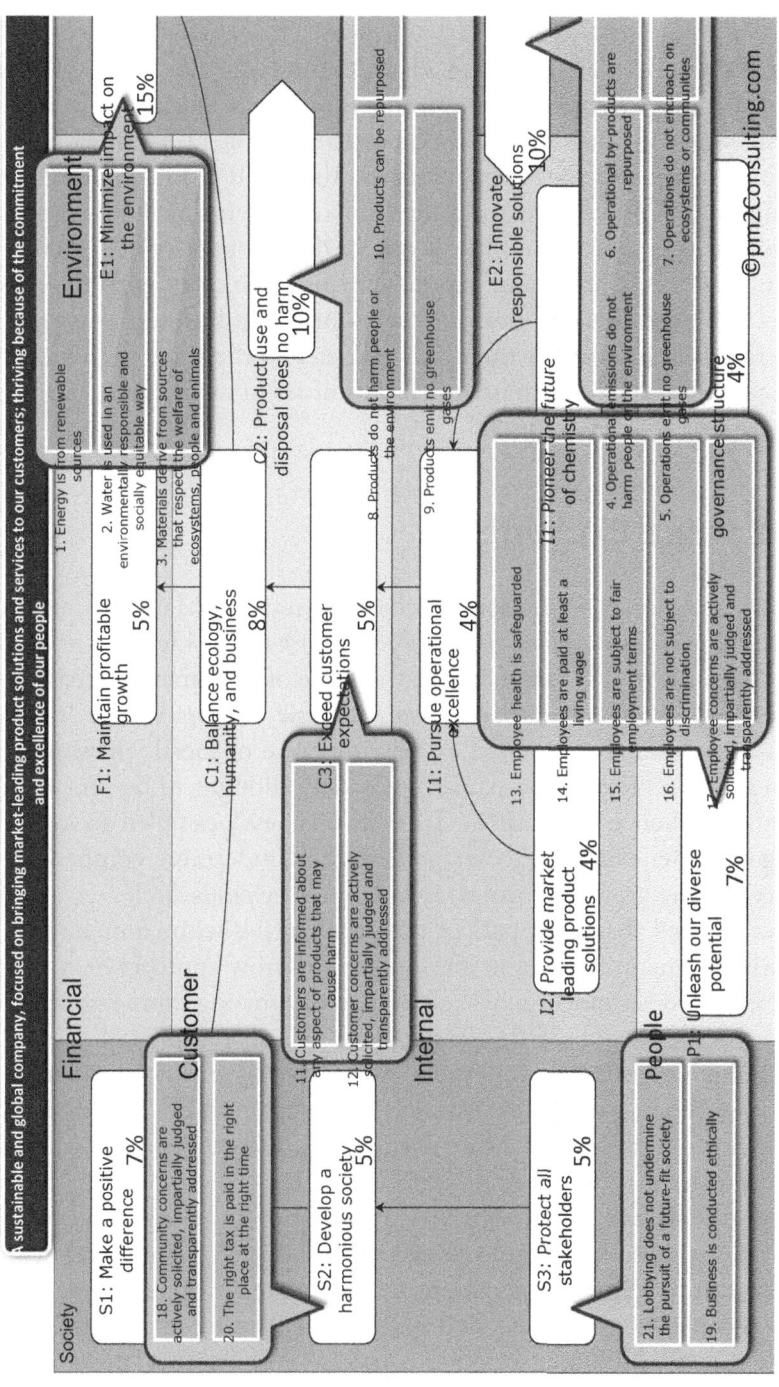

Fig. 13.2 A Sustainability Strategy Map aligned to the 21 Sustainability Business Goals. (Source: pm2Consulting)

Shared Value

Professor Robert Kaplan has been championing the use of the Balanced Scorecard to improve social and environmental performance for some time, but has an issue with the term CSR. He explains, "The problem with CSR is built into its name. 'Responsibility' suggests that CSR is a tax on companies, something they should do altruistically or to keep their stakeholder groups at bay. Viewed this way, CSR becomes a cost of doing business" [7].

Kaplan prefers a "shared value" approach which, he says, "encourages companies to think creatively to find new ways of doing business that not only generate financial returns for them but also enable them to be environmentally sustainable and to contribute to communities in which they source, produce, distribute, market, and sell."

Shared Value Explained

As a brief explanation, Professors Michael Porter and Mark Kramer introduced the shared value concept in the 2011 Harvard Business Review article, *Creating Shared Value* [8]. Launched in the wake of the financial crisis, shared value was a way to "reinvent capitalism." As the authors wrote: "In recent years, business has been criticized as a major cause of social, environmental, and economic problems. Companies are widely thought to be prospering at the expense of their communities. Trust in business has fallen to new lows, leading government officials to set policies that undermine competitiveness and sap economic growth. Business is caught in a vicious circle."

They continued that a big part of the problem lies with companies themselves, "which remain trapped in an outdated, narrow approach to value creation," Porter and Kramer argue. "Focused on optimizing short-term financial performance, they overlook the greatest unmet needs in the market as well as broader influences on their long-term success. Why else would companies ignore the well-being of their customers, the depletion of natural resources vital to their businesses, the viability of suppliers, and the economic distress of the communities in which they produce and sell?"

In essence, shared value "reinvents capitalism," by tying business value (profit, etc.) to value for communities and without environmental degradation. According to Porter and Kramer, "firms can do this in three distinct ways: by reconceiving products and markets, redefining productivity in the value chain, and building supportive industry clusters at the company's location."

As detailed in Panel 1, Professor Kaplan would like to see many more Strategy Maps and scorecards reflecting environmental and community objectives, but it's possible to take the shared value philosophy even further and build in the idea of collaboration and networks. Moreover, he explains how this all ties back to a sense of purpose.

Positive Impact

With a corporate mission to deliver a "Positive Impact"—that is social, economic, and environmental value—The Palladium Group (a firm originally led by Doctors Kaplan and Norton) is building on the shared value ideas by playing the role of a catalyst in "Positive Impact Partnerships." These partnerships work to share governance mechanisms to ensure transparency, engagement, and alignment, while holding actors to account. Through the Balanced Scorecard concept, it encourages multiple companies to co-create a Strategy Map and scorecard related to a specific partnership, with the goal to promote collaboration, alignment, and commitment among all stakeholders.

Palladium is applying this concept in its work with Syngenta, a large agribusiness, where it is supporting the development of system Strategy Maps with cascading scorecards for each of the actors involved in their good growth plan interventions in Indonesia and Nicaragua. According to Juan Gonzalez-Valero, Head of Public Policy and Sustainability at Syngenta, "If you take agriculture…it's [about]… understanding what is actually necessary to capture value before the farm, on the farm, and after the farm. If we agree on the Strategy Map overall, we can agree on certain metrics that will help all the players get involved, and I think that holds true for all sectors" [9].

Networked Organizations

Such networked organizations represent an emerging and powerful application of the Balanced Scorecard System. These networks bring together distinct organizations and groups of people to deliver collective outcomes while also achieving the goals of each individual network member. Such networked—and oftentimes virtual—collaborative structures are becoming more commonplace as organizations struggle with the complex challenges of the twenty-first century.

Networked organizations vary widely depending upon their collective goals. They might consist of public sector organizations, private sector organizations, or a combination of both. They might be short term in nature (e.g. delivering a specific innovation project) or longer term (dealing with deeply ingrained socioeconomic challenges). The structure of these collaborations can exaggerate the same issues that organizations face individually. Networked organizations struggle to assign accountabilities, maintain transparency, and measure progress. Further, the looseness of the affiliation between network members creates a culture in which it is easy to assign blame elsewhere.

The Balanced Scorecard system addresses these obstacles by aligning the disparate network members around a common strategy to which they all contribute and by creating a platform through which each partner can understand their progress toward a common goal. The following case example showcases a successful implementation of the Balanced Scorecard system across a large network of partners seeking to drive positive outcomes in their community.

Case Illustration: Thriving Weld

Thriving Weld is a collaborative effort by the North Colorado Health Alliance (NCHA) to facilitate data sharing and collaboration in Weld County, Colorado, a community of 275,000 people in the western USA. This network of over 60 partners capitalized on the Balanced Scorecard to share specific information about their plans and progress and collectively improve the community's wellbeing. Thriving Weld has built more than 40 scorecards across seven focus areas:

- Healthy Eating;
- Active Living;
- Healthy Mind and Spirit;
- Access to Care;
- Livelihood;
- Education; and
- Health Equity.

At the outset, the scorecard system appealed to the Weld County Department of Public Health and Environment because it recognized that no single organization could successfully address all the priority health issues and social determinants of health in their community. Previously, each organization developed its strategy, measurements, reporting, and communication structures independent of one another. As a result, Weld County's many efforts to improve the health and wellbeing of their community faced problems of fragmentation, wasteful redundancy, and inefficiency.

NCHA partnered with the US-based consultancy Insightformation, founded by Bill Barberg (a pioneer in collaborative scorecards), to create a community-based collaborative scorecard approach that would suit their networked environment and shift thinking from organization-centred strategies to community collaboration. Thriving Weld tailored the classic Kaplan-Norton Balanced Scorecard model to their needs, using a three-perspective structure for their strategy maps and scorecards, instead of the usual four (Fig. 13.3).

- Outcomes are what the multi-agency partnership wishes to achieve collectively. For example, the ultimate outcome in the Healthy Eating focus area is "Increased People Living at a Healthy Body Weight."
- Strategies (such as "Increase Healthy Food Options in Restaurants and Retailers") deliver the outcomes, essentially combining the conventional customer and process perspectives.
- Assets and capacity development capture the prerequisites for achieving the strategies, much as the learning and growth perspective supports the Process perspective in conventional scorecards. Here, we find objectives such as "Gather and Share Data to Improve Prioritization and Monitoring."

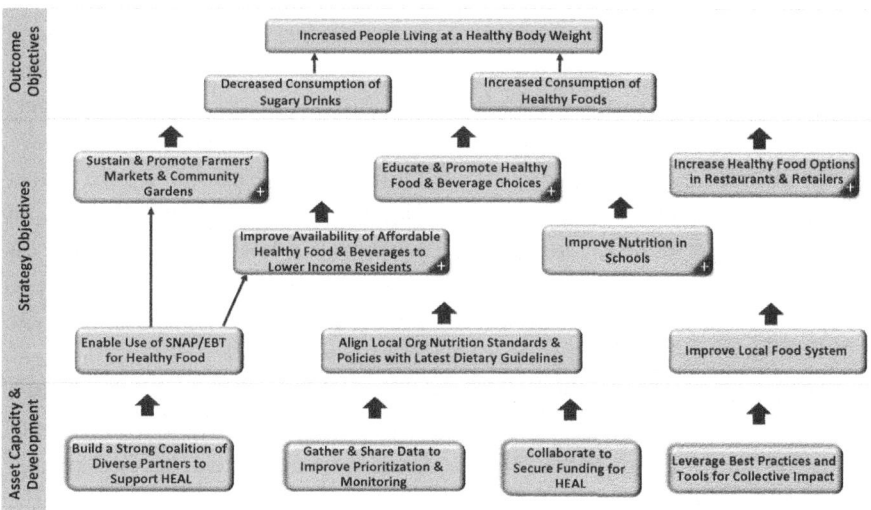

Fig. 13.3 Thriving Weld County Strategy Map. (Source: Insightformation)

As with a conventional strategy map or scorecard, cause and effect build from the bottom to the top.

Zoomable Strategy Maps

High-level focus area Strategy Maps are composed of objectives that are too big for any one organization to be solely accountable. Thriving Weld addressed this issue using technology-enabled Strategy Maps that can zoom in or out to show various levels of detail, like a digital map on a GPS. Stakeholders can view a high-level Strategy Map to understand the big picture, then zoom in and view the details of a given objective, narrowing in on those that are most relevant to them. The zoom feature enables all partners to work towards the focus area outcomes through their own spheres of work, while benefiting from the insights and data provided by the group as a whole.

Robust Shared Measurement System

The focus area Strategy Maps provide the structure for a shared measurement system. Each outcome's objective has one or two measures to monitor the community's progress.

Because changes at a community level take years to achieve, outcome measures move very slowly. Instead, progress in the short term is evident in the Strategies perspective, which supports the outcome perspective. The Balanced Scorecard methodology emphasizes the importance of leading indicators, which in this case is accomplished with driver measures and associated targets that monitor the progress of objectives in the strategies perspective.

Before launching Thriving Weld, individual organizations would select their own measures for their programs, frustrating collaborative efforts even when organizations were working toward common goals. Today, the process is inverted: organizations make program and funding decisions according to their expected impact on the driver measures, which are shared across the network—providing a powerful incentive for teamwork.

As with a conventional scorecard system, thoughtfully selected initiatives propel progress on objectives—but Weld County uses the term action instead. Community-oriented organizations are accustomed to creating and implementing action plans, so the familiar term "action" resonates better than the more nebulous "initiative." Language is important for buy-in.

Transparency and visibility, also central to buy-in, are seen as the keys to success. The Strategy Maps and scorecards for dozens of community partners are available for the public to view on Thriving Weld's website [10]. These pages automatically update with the latest data from the centralized measure collection system.

The Million Dollar Question: Is This Approach Creating the Desired Impact?

When dealing with deeply entrenched community issues like obesity and chronic disease, the time lag between embracing a new approach and seeing the outcome in the community can be considerable. It takes time to change the systems, the environment, and people's behaviours. It takes time for those changes to impact health outcomes. It takes time to collect and validate data.

Although it is early still to see the kinds of results Thriving Weld eventually intends to produce, there are, however, many encouraging signs of progress that show that this approach is creating alignment, action, and positive movement to implement the strategy.

Improved Communication

A year into Thriving Weld's Balanced Scorecard implementation, Dr. Mark Wallace, Weld County Health Department's Executive Director, gave a presentation at a public health conference. He told the audience,

> We've had a lot of energy and activity around becoming a healthy county for quite a few years. But in the past, it was like all the organizations were…people talking in a crowded restaurant. All the different conversations ended up creating a lot of noise. Now, with a shared strategy that is clearly communicated using the strategy maps, it is like being in a theatre with surround sound [11].

Improved Alignment of Funders and Other Stakeholders

Grant makers have a huge impact on how non-profit organizations work in a community. Because of their participation in Thriving Weld's Balanced Scorecard program, the United Way of Weld County decided to align their funding to the objectives on the Strategy Maps and have grantees report their performance using the shared measurement system. In turn, this decision led to greater interest among community organizations in how Strategy Maps can support improvements to critical goals.

Improved Engagement and New Actions

Having a clear set of shared priority objectives has inspired many community organizations to launch creative actions to address their community's needs. This list of new efforts is long and varied. For example, in 2015, community partners launched a new initiative to improve active recreation for minority youth by offering "Family Fun Passes" to a nearby funplex. Nearly 1400 families registered and over 3500 passes were used. In 2014, 35 organizations incorporated plans or messaging about active living and reducing the consumption of sugar-sweetened beverages into their communication and modelling to youth. While many of these initiatives are small, their collective impact reveals the strength of the community's engagement in achieving big goals together.

Parting Words

The Thriving Weld experience and other similar ventures (see, as examples, the Insightformation or Palladium websites, [12, 13]), show how collaborative scorecards can address complex social challenges, from improving the quality of life in economically disadvantaged communities to addressing crisis areas such as drug abuse. The potential is far-reaching—by adapting a system developed for the private sector to address the ever-evolving needs of the public sector, organizations spur innovative solutions to truly pressing problems. Such examples provide compelling evidence of delivering positive collective impact through a collaborative, networked structure.

Professor Kaplan provides this powerful statement, which can serve as a call to arms! "In his Gettysburg Address, President Lincoln spoke about democracy as 'government of the people, by the people and for the people.' Shared Value strategies today are about delivering value for citizens. A collaborative and co-created Shared Value Strategy can channel Abraham Lincoln by having the shared value created 'of the people, by the people, and for the people.'"

Panel 1: Co-Created Balanced Scorecard

In a June 2015 edition of Strategically Speaking (which was edited by one of the authors of this book), selected experts were posed the question, *"How Can Companies Do More Good for Society and Turn a Profit?"* The following is an extract from the reply provided by Professor Robert Kaplan, [7].

A Changing Social Context
With the growth of democracy around the globe during the past 50 years, more people want their voices to be heard. They voice their concerns as to the adverse social and environmental impacts from companies working within their local communities.... people are rebelling against the massive pollution being created by power generation and industrial operations. After not paying attention to these externalities for 200 years, companies should be able to find many opportunities ("low-hanging fruit") to respond to these concerns without having an adverse effect on their bottom line.

Beyond the spread of democracy, the Internet and social media have given people much more knowledge about standards of living and working practices in various parts of the world. This information causes them to question the practices of companies in their communities and nations.

As people in developed nations move up the Maslovian scale of needs, they no longer worry about basic needs such as food and shelter. Many seek more fulfilling and meaningful lives. They want to work for organizations that can help deliver that higher sense of purpose work that they feel good about and whose mission they believe in and aligns to their own.

Shared value helps companies meet employees' expectations for work that not only provides employment and income, but also delivers value to all a company's stakeholders—including, of course, its shareholders. A company creates meaningful employment when it delivers on its mission, defined by how it creates value for customers. The mission statement of the pharmaceutical company Merck, for example, says that medicine is for people, not profits. But, Merck also notes that the better they have gotten at delivering medicine for people, the more profits have followed. Profit is a score of how well a company delivers value to its customers.

The Balanced Scorecard
Dave Norton and I developed the Balanced Scorecard because we recognized that quarterly and annual profit are not a complete metric of whether a company created or destroyed shareholder value in a given period. The way to maximize long-term value is to continue to be innovative in delivering value to customers. Innovation is also central to a good shared value strategy. Companies must search for new and better ways to create value, sustainably, for shareholders, customers, suppliers, employees, and communities throughout their value chain.

Over the past 20 years, we have interacted with many companies that want to quantify their performance in making positive social and environmental impacts. Several have created environmental and community performance strategic objectives on their maps and several have introduced a fifth perspective for environmental and social performance.

This made such performance a core part of the strategy for which the entire executive team would be accountable. In contrast, the CSR approach was often delegated to a separate function or the corporate foundation.

Co-Creation
I would like to see many more strategy maps and scorecards reflecting environmental and community objectives. But, we can take the shared value philosophy even further.

> The best shared value strategies should be co-created by companies, NGOs, community representatives, and other key stakeholders. The co-creation creates a consensus as to the value that must be delivered for each group and who will be responsible and accountable for the various objectives.
>
> The co-creation process also creates trust and understanding among the groups. The non-corporate stakeholders will recognize that a corporate commitment to the shared value strategy can only be sustained if the company earns adequate financial returns from the strategy. Even after the shared value strategy map and scorecard have been co-created, the collaboration process continues by inviting the stakeholders to participate in strategy review meetings to discuss performance, accomplishments, shortfalls, and new improvement opportunities. A good measurement system will be critical, and all parties will need to build robust metrics for environmental, social, and community impacts.

Self-Assessment Checklist

The following self-assessment assists the reader in identifying strengths and opportunities for improvement against the key performance dimension that we consider critical for succeeding with strategy management in the digital age.

For each question, any degree of agreement to the statement closer to one represents a significant opportunity for improvement (Table 13.1).

Table 13.1 Self-assessment checklist

Please tick the number that is the closest to the statement with which you agree		
	7 6 5 4 3 2 1	
My organization has a very strong sense of social and environmental responsibility		My organization has a very weak sense of social and environmental responsibility
My organization has very well defined social and environmental strategic objectives		My organization has not defined any social and environmental strategic objectives
My organization has a very effective process for managing external networks		My organization has a very ineffective process for managing external networks
My organization has a very effective process for managing internal networks		My organization has a very ineffective process for managing internal networks

References

1. Archie Carroll, *A History of Corporate Social Responsibility: Concepts and Practices*. www.academia.edu, 2013
2. *Social Responsibilities of the Businessman*, Harper, 1953
3. James Creelman, *Creating a Balanced Scorecard*, Lafferty Publications, 2001
4. See, Novo Nordisk – *Managing Using the Triple Bottom Line Business Principles*, CSR Europe, 2013
5. *Novo Nordisk 2016 Annual Report*, Novo Nordisk
6. Brett Knowles, *The Sustainability Strategy Map*, Linkedin blogs 2017
7. Robert S. Kaplan, *How Can Companies Do More Good for Society and Turn a Profit?* Palladium, June 2015
8. Michael Porter and Mark Kramer, *Creating Shared Value*, Harvard Business Review, January/February. 2011
9. See http://thepalladiumgroup.com/research-impact/Governing-Positive-Impact-Partnerships--laying-the-foundations-for-success
10. See http://thrivingweld.com/
11. James Creelman, Bill Barberg, Mark E. Wallace, *Driving Positive Community Outcomes with the Balanced Scorecard: Collaborative Strategy Management in a Community Network*, Palladium white paper, 2015.
12. www.insightformation.com
13. www.palladium.com

14

Conclusion and 25 Key Strategic Questions

Introduction

Strategy management is evolving and will continue to evolve—as will how organizations are structured, processes are optimized, and value is delivered to the customer, as well as the dynamics of the employer/employee relationship. When we passed through the technology "tipping point" and entered the unknown territory currently called the digital age (a description that will also evolve), we had no idea what would change—except one thing: everything.

If everything changes, but we're rarely sure what, when, or how (although, as we explained in Chap. 2, approaches such as technology-based planning can help here) then organizations, in any sector or industry, have to develop, and as a core competence, build exceptional capabilities around agility and adaptiveness.

Agile and Adaptive

As we stated in the opening chapter, agile points to sudden quick changes—being "able to move quickly and easily," whereas being adaptive is about "having the ability to change to meet different circumstances" (which does not necessarily mean quickly or easily). Both are required.

Building on the earlier work of pioneering thinkers, most notably Doctors Robert Kaplan and David Norton, in helping organizations transition into what was known in the 1990s as the knowledge age, this book has outlined a model that helps enable organizations to be more agile and adaptive in the management of strategy in today's digitally driven, knowledge-based environment.

25 Key Strategic Questions

To aid the reader in implementing the agile and adaptive principles described in the model, we have set the following 25 questions, the answering of which should trigger thoughts regarding the challenges to be overcome and the capability developments required. Questions have been collocated according to the five stages of the model, as well as the leadership and cultural and people underpinnings.

Stage 1: How to Formulate Strategies for the Digital Age

1. What does the word "strategy" mean to your organization?

Strategy has many definitions. Although there's no perfect description to prescribe, it is important that there's a common agreement about the term among the executive leadership team. Without agreement, little chance do they have of being certain of shaping the most appropriate objectives or initiatives. And how can buy-in possibly be achieved? If the senior team doesn't know what strategy means to the organization, how can anyone else?

Task
Ask each member of the senior team to write down on a piece of paper what he or she understands by the term "strategy." Then ask each to read out their definition. The difference in interpretation can be astounding. Finally debate, discuss, and agree upon what strategy means for the organization.

2. To what extent is silo-based working stymieing organizational performance?

Sadly, most organizations are still structured along the lines of the strict silo-based working diktats of Frederick W. Taylor's *Principles of Scientific Management* [1]. This has led to internal departmental turf-wars, and often has a detrimental effect on inter-departmental team working. This will be a major performance blocker in the fast-paced digital age.

Task

Consider designing and implementing end-to-end process management, with strategic processes led by a member of the executive team. The expert work of silos still gets done, but in the context of ultimate process outcomes, as opposed to departmental objectives.

End-to-end process management should certainly be applied to strategy. The focus should be on recasting strategy management as a single, integrated, end-to-end process rather that the siloed approach through which leaders and strategic planners create the strategic plan and then hand over to managers to execute "as is."

3. What is the "function" that the organization delivers?

As far back as the early twentieth century, pioneer thinkers such as Henry Ford recognized that customers buy a "function" that fulfils a specific need, not a product. People did not buy coaches and horses; they bought a mechanism for travelling quickly and comfortably: cars did this much better. Kodak did not sell film; they sold a way to capture images: the iPhone did this much more conveniently. The same disruption can be applied to any industry or sector and in the digital age and at alarming speed.

Task

Get the senior team to write down the answers to following questions and make it clear that very different responses are required:

- What products or services do we sell
- What do we sell?

4. Is the organization clear as to its "sense of purpose?"

The mission statement defines the purpose of the organization: why it exists. A mission generally doesn't (or at least shouldn't) change much over time. Google's mission "To organize the world's information and make it universally accessible and useful" has been stable since the firm's launch in 1998.

Task

The task here is simply to write the mission statement, ensuring it does not mention a specific product.

5. Has your organization crafted a quantified vision?

The vision statement describes the desired result at the end of the strategic horizon and is the anchor to the subsequent strategy development and execution phases. Unfortunately, most vision statements are generic or vague and oftentimes little more than advertising slogans.

Task
Review the vision of the organization to ensure it includes the following three components:

- It is time-bound (has an end date)
- It is measurable (for a commercial organization, this will typically be the measures of ultimate financial success)
- Captures the essence of the unique value proposition to the customer.

6. Create a Strategic Change Agenda

A useful tool for transitioning from defining the quantified vision to a Strategy Map is a Strategic Change Agenda. This is a framework to identify and assess the current states and to project desired future states for strategically critical performance dimensions.

Task
Structure performance dimensions (previously identified through one-to-one interviews with the senior team) to the Balanced Scorecard perspectives. As examples,

- Financial (profit dimension): From losing money to achieving profitability of X.
- Customer (product choice dimension): From limited number of traditional products to diversified range of innovative and value-added products and services.
- Internal Process (supplier dimension): From traditional "arms -length" relationship to partners in identifying and delivering customer solutions.
- Learning and Growth (employee dimension): From demoralized and adversarial to highly motivated and participative.

Get the change agenda right and the objectives simply fall out.

Stage 2: How to Build an Agile and Adaptive Balanced Scorecard

7. Develop objective statements

An issue with strategic objectives is that they are limited to a few words, such as "Enhance Customer Experience." Succinct, yes, but its meaning will be interpreted in various ways as it cascades down the organization. So always write a meaningful objective statement that more fully describes the meaning of the objective.

Task

- Write a "statement" for each objective within the financial and customer perspectives. This short paragraph should describe the desired strategic outcome.
- For internal process and learning and growth, add a second paragraph describing how the desired outcome will be achieved.

8. Has your organization identified the key strategic risks for each strategic objective?

Risks impact each objective on the Strategy Map—financial and non-financial. In implementing strategy, organizations need to "keep one eye on performance and one of risk." This is fast becoming a pressing requirement in the digital age.

Task

For each strategic objective, identify the risk events that could lead to failure to achieve targeted performance. Then focus on shaping plans to mitigate strategic risks, or even undertaking those deemed necessary to achieve the objective.

At the outcome level of the Strategy Map, risks should be articulated as the effect: "what might happen if the risk materializes," while, at the enabler level, "what might cause unwelcome outcomes."

9. How does your organization select KPIs?

Organizations typically over-populate their Balanced Scorecards with KPIs. This typically due to struggles to identify the most impactful measure for a strategic objective. An unwelcome outcome being that the scorecard system becomes little more than a mechanism for capturing and reporting data—and generally disliked by the bulk of employees.

Task

Use driver-based models to identify the three or four most critical "do wells" for delivering to the objective (as described in a well-designed objective statement). Then apply Key Performance Questions (KPQs) to those drivers. KPQs highlight what the organization needs to know in terms of delivering to the drivers, enabling a focused discussion on how well it is delivering to the "do wells."

10. How well have staff that work with measures been trained in the "science of measurement?"

Most organizations are obsessed with measurement, but don't invest the time and money into teaching those that work with measures even the basics of the underpinning science. We would expect a finance professional to understand finance and the same for an IT specialist, but not for those working with KPIs. We need to redress this odd and dangerous omission.

Task

The task here is to ensure that employees that regularly work with KPIs (and especially those that must comment on a performance results) have received at least basic training in how measures work.

11. When setting KPIs and associated targets, to what extent have you considered the behaviour that will be driven?

A poor understanding of the science of measurement also means that organizations often overlook the fact that measurement does not always drive the expected behaviours.

Dysfunctional behaviours (which are simply rational responses—that is, doing what is required to hit the target) triggered by a KPI are far from uncommon. It is well known for a manufacturing plant to set a target to reduce reported injuries, and for the target to be reached simply by only reporting serious injuries (that can't be hidden). Performance does not improve, but the target is hit.

Task
In a workshop setting, brainstorm and write down all the positive behaviours that might be encouraged and then the negatives. When done, hold a team discussion on how to best encourage the former and mitigate the latter. Sometimes, the risk of dysfunctional behaviour is so great that the KPI has to be rethought or abandoned.

12. How well do you understand the meaning of strategic initiatives?

As with KPIs, there are often way too many strategic initiatives on a Balanced Scorecard. This is often due to a poor understanding of the difference between a strategic initiative and a business as usual task.

Task
Ensure that each strategic initiative complies with the following.

- Has a defined start and end date
- Is a unique undertaking and a task that is repeatable
- Is important enough to require sponsorship from a member of the senior leadership team
- Has the required financial and human resources

Ideally, initiatives should only be directly linked to the internal process and learning and growth perspectives (as it is what gets done here that delivers the financial and customer outcomes). Ideally, initiatives should impact more than one strategic objective. Too many organizations focus on one-to-one links, which can be sub-optimal and very silo-focused.

Stage 3: Driving "Rapid" Enterprise Alignment

13. How well do you ensure buy-in to the strategy through the cascade process?

The conventional approach to cascading the Balanced Scorecard system is typically a lengthy and time-consuming process, top-down driven and largely imposed. This rarely leads to buy-in and is no longer fit-for-purpose in today's fast-moving markets.

Task

Identify the critical (and very few) objectives and KPIs to mandate in the cascade and then empower teams to build their own scorecard systems that describes what they want to achieve over the coming period.

As well as leading to greater buy-in and ownership, this transmits a message that senior management trusts their employees and believes in their abilities. Proper governance still ensures alignment, but guided by flexibility and empowerment instead of rigid imposition.

14. Do teams understand their own "sense of purpose?"

Firstly, when cascading, use the word strategy less and less as it is taken deeper into the organization. Top-level executives should make strategy their number 1 priority, and they can pull the levers to drive transformational change.

At deeper levels, such levers are not available so focus on the sense of purpose of the department or team—what they want to achieve over the coming period and then work on linking this to the strategic goals.

Task

When building devolved Strategy Maps and Balanced Scorecards at departmental/team levels, focus on the purpose of the group. Discuss:

- The sense of purpose of the organization as encapsulated in the mission
- How the group relates with other parts of the organization
- The individual's own sense of purpose: what they want to achieve over the short and medium terms and when, why, and how.

Stage 4: Getting Results Through Agile Strategy Execution

15. To what extent are your Strategy Office (OSM) and Project Management Office (PMO) collaborating?

We have encountered some organizations in which the PMO and OSM refuse to talk to each other! As well as leading to a sub-optimized initiative management process, this invariably means that beleaguered managers in the field have to report performance on the same project/initiative through two different systems—a strategy management system and project management information system. A further dysfunctional outcome is that the PMO and the OSM interpret, and so report, the results in oftentimes very different ways, to the continued ire of the senior team.

Task

Consider merging the OSM and the PMO, as both are part of a single process that delivers strategic initiatives.

Even so, there is still the question of who does what in the initiative management process? The PMO's expertise is around ensuring that projects are managed efficiently and to established procedures—be they strategic initiatives or large tactical or operational projects.

An OSM is primarily required to manage the strategy process and facilitate strategic alignment as well as the review and update of the strategy. It is not the PMO's job to make the strategy link, just as it is not the OSM's job to manage projects (tactical or strategic). Clear delineation of roles is required, even when capabilities are merged—so write a charter describing the roles and responsibilities of both.

16. To what extent are financial management processes aligned to strategy?

Professor Kaplan once said that, "One aspect of the Strategy-Focused Organization that has lagged is the integration with the budgeting system… if we don't establish the link with budgeting, then scorecard initiatives may wither." [2]

Quite simply, the budgeting process should support the strategy management process. Indeed, sequentially, it should be completed after the strategic plan and mid-term plan.

Task

Consider de-emphasizing the budget. Rather than a fixed performance contract, the budget should capture stake in the ground annual targets. Based on mid-term strategic goals, these should be stretching—what is possible, not most likely. A rolling forecast, looking perhaps 4–6 quarters ahead (with greater detail over earlier quarters) and based on the most up-to-date information and insights, becomes the main process for steering the organization's finances.

17. How well has your organization identified the operational drivers of strategic objectives?

Too many organizations believe that linking operations with strategy is about populating department level scorecards with operational KPIs. This is a misunderstanding, and contributes to scorecard systems being a mass of measures.

Connecting strategy and operations is done via the objectives within the internal process perspective of a Balanced Scorecard system.

Task
Use driver-based models (which asks the question "what must we excel at to deliver to this objective?") to identify the key operational processes that support the objectives within the internal process perspective of the Strategy Map. With the drivers identified, then identify the measures for these sub-processes and monitor them through an operational dashboard.

Stage 5: Unleashing the Power of Analytics for Strategic Learning and Adapting

18. How well does your organization understand "cause and effect?"

Most Strategy Maps that we review are not maps but collections of, at best, loosely related objectives arranged according to four perspectives. Useful for communication and alignment, but not exploiting the real benefits of a Strategy Map—causal analytics. Next generation maps, and overall scorecard systems, will drive performance by leveraging advanced data analytics: testing the hypotheses that inform the choice of objectives and KPIs, as well as the impact of initiatives and other improvement interventions.

David Norton refers to this as the next evolution of the Balanced Scorecard system, adding that excellent analytics capabilities are becoming a "must have" capability for strategy offices or OSMs.

The *task* here is simply to start building an advanced data analytics capability.

19. How well does your organization understand that a KPI is an indicator of performance, not the whole story?

For the term Key Performance Indicator, the word indicator is important. It is an indicator of performance, not the complete answer. When measuring strategy, we are not seeking absolute measures of performance, as we might when measuring operational processes.

Task
When reviewing KPI "scores," always ask what is it not telling me? Also, keep in mind that the top-level score is not particularly meaningful without drilling into the supporting data.

20. How well does the organization understand the monitoring performance is not just through KPI scores?

Periodically, when preparing the report for the strategy review meeting, print these words from Albert Einstein on the front cover. "Not everything that counts can be counted and not everything that can be counted counts."

This occasional task serves as gentle reminder that the meeting should be a conversation on performance and not a discussion on KPIs, which are simply an input into the discussion.

21. To what extent is the annual strategy refresh still of value to your organization?

Many commentators argue that the annual strategy refresh has passed its sell-by date. That it is no longer acceptable to manage strategy through a process where it is reviewed once a year and then frozen until the next annual update.

Task
The management team should review how well the organization captures external market/customer moves and is able to synchronize the internal and external rates of change. Strategic moves should be triggered by an external event (competitor or customer move or, better still, early identification of a trend). At the same time, there has to be clarity around the firm's strategic choices and positioning (which remain longer term).

Underpinning the Model

How to Ensure a Strategy-Aligned Leadership and Culture

22. How well does your organization understand the impact of leadership behaviour on culture?

Leadership and culture are essentially indivisible. Culture is demonstrably shaped by the leadership team (for better or worse). The alignment of leadership and culture is an imperative for strategies to succeed and for organizations to be sustainable.

Task
A useful technique is to think of the shadow of the leader. This shadow casts a long way within an organization, from which employees know (and therefore do) what the leaders really want and reward—not what the leaders might say they want, and even espouse in public statements, internal communications, or even corporate values.

23. Values are more than just Behaviours

Making values stick requires changes to the way the organization operates. If an organization has a value around trust, then it is inappropriate to require lower level managers to require multiple sign-offs before spending relatively small amounts of money. If no significant changes are made, the values will be no more than "nice sounding words," that no one can argue with, but equally will take no notice of.

Task
When designing/agreeing values with the executive team, ask them how driving these behaviours will be supported by significant changes to the following: policies, procedures, processes, information flows, decision-making, metrics, and incentives. If these are not changed, then the values will remain "nice sounding words."

24. Cultural Assessment

A cultural assessment can provide early warning signs of cultural barriers to strategy execution. Culture can ultimately derail even the most beautifully crafted strategic plan.

A cultural assessment will show in which parts of the organization resistance might be fierce and which might be more accepting—and so launch appropriate and targeted initiatives or actions.

Task
A powerful way to effectively analyse culture is through an online cultural assessment tool, which will reveal the areas where improvement is needed enterprise-wide or at departmental levels. The latest generation of online tools uses advanced data analytics to more precisely assess the organizational culture and provide improvement recommendations.

Creating a Strategy-Aligned Workforce for 4th Industrial Revolution

25. How well is the sense of purpose of the individual aligned with that of the organization?

For the individual employee, sense of purpose has two dimensions. The first is about what they want to achieve professionally, and while with the present organization. The second dimension, which is increasingly evident in Millennials and Generation Z, is the desire to work for an organization that shares their commitment to being better citizens.

Task
Think about how well the organization aligns the individual's sense of purpose with that of the enterprise. This must be central to the recruiting and bringing-on-board process. Organizations must be honest and state that they expect the individual to stay for as long as s/he and the organization gain value (and ensuring there's a strong emphasis on the individual) and that the sense of purposes remain aligned (for this, the organizational sense of purpose must be clear).

Personal development plans should start from what the individual wishes to achieve/learn over the next period (both the short and longer term) and mapping this to the goals/capability requirements of the enterprise (if there's a poor match, then there's a red flag to address).

Final Words

Peter Drucker once said that "the best way to predict the future is to create it"—although this is often attributed to others, including Abraham Lincoln [3]. Whoever did say it, said it many decades ago; but this is profound advice as we sequence through the early years of this the 4th industrial revolution.

However, powerful advanced data analytic tools are enabling more reliable "peering" into what the future might look like and bringing these observations into the present. To an extent, we are "co-creating" the present and future at the same time, while drawing on lessons from the past. Albert Einstein once wrote, "…the distinction between past, present and future is only a stubbornly persistent illusion" [4]. As the management of strategy continues to evolve, perhaps we can say the same about how we understand its formulation, execution, and learning.

References

1. Frederick W. Taylor, *The Principles of Scientific Management*, Harper and Brothers, 1911.
2. Lori Colabro, *On Balamce*, CFO Magazine, February 2001
3. Peter Drucker *Attrib*.
4. Alice Calaprice, ed., *Einstein, The Expanded Quotable Einstein*, Princeton: Princeton University Press, 2

Index

A

Advanced data analytics, 8, 10, 17, 20, 79, 80, 167, 169, 173, 175, 186–188, 204, 216, 222, 268, 271, 272
Age 2 Balanced Scorecard systems, 3, 19, 20, 80
Agile and Adaptive Model for Strategy Management in the Digital Age, 5, 14–18
Agile Manifesto, The, 23, 24, 153
Alibaba, 39–40
Alignment: Using the Balanced Scorecard to Create Corporate Synergies, 123
Art of War, The, 12, 65
AW Rostamani, 77, 215

B

Balanced Scorecard, 2, 3, 5, 16, 18, 26, 48, 55, 56, 60, 61, 80, 83–86, 89–111, 114–118, 120, 121, 126, 134, 137, 140, 144, 147, 151, 167, 172, 173, 179, 180, 183, 188, 207, 212, 214–215, 217, 218, 244–246, 248–254, 262–266

Balanced Scorecard System, 3, 5, 6, 15, 16, 47, 53, 56, 63, 69, 70, 72, 79, 80, 84, 85, 104, 105, 109, 113, 115, 117, 122, 126, 129, 131, 151, 169, 182, 214, 232, 245, 249, 250, 265, 268
Barberg, Bill, 251
Beyond Budgeting Institute, The, 129, 131, 135, 139
Big Data analytics, 170–172, 177
Blockbuster, 30, 31, 195
Blue Ocean strategy, 14, 34–35
Bogsnes, Bjarte, 97, 98, 117–119, 121, 131–134, 218–220
Boundary-less working, 11
Boyd, John, 64
Business Model Canvas, 36
Business model innovation, 14, 23, 32

C

Cascade process, 115–117, 119, 120, 122, 125, 126, 139, 140, 144, 163, 174, 218, 249, 263, 265–266

Cause-and-effect relationships, 10, 56, 74, 78–83, 119, 181, 183, 189, 252, 268–269
Choosing strategic initiatives, 106–109
Christensen, Clayton M., 12
Churchill, Winston, 10, 192–194, 205
Cigna Property & Casualty, 47, 48, 80
Cirque du Soleil Coffey, Jim, 34, 35
Collaborative scorecards, 18, 243–256
Corporate values, 191, 210–211, 213, 214, 217, 270
Cox, Jeremy, 103, 104
Customer co-creation, 14, 36, 37

D

Deming, W. Edwards, 114, 141, 229–231
Descriptive analytics, 171–175
de Vries, Andreas, 147, 207, 209, 210
Disruptive innovation, 12–13, 23, 29, 32–33, 42
Driver-based models, 15, 16, 93, 136, 145, 264, 268
Drucker, Peter, 207, 272
Durham Constabulary, 94–96
Dysfunctional behaviours, 101, 102, 105, 208, 216, 238, 264, 265

E

Einstein, Albert, 34, 70, 92, 126, 183, 269, 272
Eisenhower, Dwight, 19
Emotional touchpoints, 38–39, 43, 179
End-to-end process management, 10–12, 74, 77, 93, 261
Enterprise Synergy Model, 123–125
Essex Police, 236–237

Evans, Jade, 193, 197–200, 231
Execution premium process (XPP), 2, 4, 5, 13, 139, 140, 151, 152

F

Federal Bureau of Investigation (FBI), 60–62
Finance-based planning, 33–36, 41–42
Five Forces, 9, 49, 50
Ford, Henry, 29
Fordism, 6
Ford Motor Company, 6, 7, 12, 29, 31, 33, 39
Forecasting, 16, 100, 105–106, 120, 131–133, 135–139, 172, 182, 183, 188, 267

G

Gallup, 7, 225–226, 234
Gartner, 50–52, 171
Global Center for Digital Business Transformation, 201, 204
Google, 28, 29, 142, 191, 219, 225, 228, 232, 261

H

Hackett Group, The, 13, 124, 125, 135, 144
Highsmith, Jim, 24
Hope, Jeremy, 82
Hoshin Kanri, 113–115

I

Intellectual capital model, 82–83
Ionescu, Mihai, 12, 13, 49, 90, 96, 123

J

Jobs, Steve, 33, 39, 192–193

K

Kaplan, Robert, viii, 2, 3, 5, 15, 60, 61, 70, 83, 85, 104, 107, 110, 116, 123, 125, 129, 134, 139, 140, 147, 151, 155, 167, 179, 183, 248, 249, 254, 259, 267

Key performance indicators (KPI), 7, 15, 16, 20, 49, 56, 57, 61, 69, 74, 78, 85, 87, 89–103, 105–107, 109, 110, 113, 116–118, 120, 122, 126, 131, 137, 145, 155, 159, 160, 163, 169, 170, 172, 173, 176–178, 180–183, 185, 188, 214, 215, 229–230, 238, 245, 263–266, 268, 269

Key performance questions (KPQ), 15, 92–96, 264

Kin, W. Chan, 34

KiwiBank, 181–182

Knowles, Brett, 246

Kodak, 29–31, 143, 195, 209, 228, 261

Korea South-East Power (KOSEP), 212–213

L

Leadership for the execution of strategy model (LFES), 193–199

Leadership Forum Inc., The, 9, 184

M

Marr, Bernard, 92

Mid-term visions, 47–48, 66, 120, 136

Mintzberg, Henry, 9

Mission statement, 14, 28–29, 45, 231, 255, 261

Motorola, 141–143

N

Netflix, 30, 31

Norton, David, 2, 3, 5, 15, 19, 70, 75, 80, 83, 107, 116, 123, 125, 139, 140, 147, 151, 155, 160, 167, 169, 179, 183, 199, 237, 249, 255, 259, 268

Nova Nordisk, 244–245

O

Objective statements, 15, 73–75, 79, 90, 92, 93, 145, 215, 239, 263, 264

Office of Strategy Management (OSM), 89, 105, 155, 156, 158, 162–163, 172, 174–177, 182, 185–188, 266–268

OODA Loop, The, 14, 64, 65

Operational dashboards, 144–146, 174, 179, 268

P

Palladium Group, 31, 107, 193, 198, 199, 208, 230, 249

Pattern-based strategy, 50–53

PESTEL, 9, 49, 50

Porter, Michael, 27, 40, 49, 248

Predictive analytics, 172–175

Project Management Institute (PMI), 153, 156, 164, 165

Q

Quantified vision, 10, 14, 47–49, 55, 62, 69, 104–106, 120, 136, 159, 261, 262

R

Richardson, Sandy, 78

Rolling forecasts, 16, 105, 132, 135–138, 183, 188, 267

Ryan, Liz, 228
Ryder, Don, 138

S

Saatchi & Saatchi, 117, 119, 126
Saint-Onge, Hubert, 10, 70, 82, 227
Schneider, Alistair, 205
SCOPE, 53–54, 61, 153, 159
SCRUM, 153–155
Senior management interviews, 55–59, 69, 70
Sense of purpose, 14, 18, 23, 28–33, 45, 104, 120, 122, 124, 205, 225–241, 245, 249, 255, 261, 266, 271
Shadow of the leader, 213–214, 240, 270
Simpson's Paradox, 99–100, 102
Six Sigma, 115, 141–144
Sprint methodology, 24–25
Statoil, 97, 117–120, 131–135, 211, 218–220, 244
Strategic change agenda, 14, 49, 55, 57–63, 69, 70, 80, 122, 262
Strategic initiatives, 48, 56, 62, 77, 84, 90, 106–110, 115, 147, 151, 155–157, 159, 160, 162–165, 188, 265, 267
Strategic risk management, 110
Strategic targets, 15, 115, 122, 188
Strategic themes, 11, 61, 75–79, 85, 109, 181, 215
Strategy map/mapping, viii, 2, 3, 10, 11, 15, 16, 18, 20, 36, 55, 56, 59, 61–63, 69–90, 95, 96, 107, 108, 114–116, 121, 123, 124, 126, 140, 144, 146, 151, 163, 167, 173, 175, 179–181, 183, 188, 207, 214, 236, 243–256, 262, 263, 266, 268
Strengths, weaknesses, opportunities, and threats (SWOT), 9, 49, 51–55
Sun Tzu, 12, 64–65
SurveyTelligence, 220, 221

T

Taylor, Frederick W., 6, 7, 9, 11, 16, 26, 78, 141, 220, 231, 260
Taylorism, 7, 11, 134
Technology-based planning (TBP), 14, 33, 34, 39, 41–42, 259
Thai Carbon Black (TCB), 114–115
Theme teams, 77–78, 181, 182
Theory X management styles, 219, 220, 231, 234
Theory Y management styles, 219, 231
Thriving Weld, 250–254
Transformation Office, 156, 158, 159
Trend analysis, 100
Triple Bottom Line, 244, 245
Turkcell Superonline, 158, 162–164

V

Value gap, 23, 48–49, 55, 62, 106, 159–161
Vision statement, 14, 45–48, 262
Voice of the customer, 36–42

W

War games, 63–64
Wicking, Iain, 33, 42

The manufacturer's authorised representative in the EU is Springer Nature Customer Service Centre GmbH, Europaplatz 3, 69115 Heidelberg, Germany. If you have any concerns regarding our products, please contact ProductSafety@springernature.com

Printed and bound by CPI Group (UK) Ltd, Croydon, CR0 4YY

23/03/2026

02076735-0019